# Negotiating for Water Resources



Over 90 per cent of the world's population lives in countries that share a river basin with others. Freshwater resources are scarce and different nations, actors and users compete for limited rescurces in transboundary river basins; often conflicting with each other. Water is a resource with no substitute: it cannot be secured in sufficiently large quantities through long-distance trade deals; and, due to the interconnectivity of the hydrological system, the actions of one country in its water management have a direct bearing on the interests of neighbouring countries. For instance, in the Mekong River Basin, current hydropower and navigation developments in certain countries impact on traditional sources of income such as fisheries, and rice production in others. These kinds of changes in water use have given rise to conflict between countries in that region and others, but have also led, in some cases, to greater cooperation. The past few decades have seen a number of new agreements about the sharing of river resources and cooperation between riparian states.

*Negotiating for Water Resources* explores the drivers of conflict and cooperation between states in transnational river basins. Drawing on extensive fieldwork and interviews on the Mekong, Danube and La Plata river basins, the book provides a three-level analysis across three case studies, including the regional framework (EU, ASEAN and Mercosur), the River Basin Organisations (ICPDR, MRC and CIC) and the micro level. The key question of the book is: To what extent do power asymmetries prevent or inhibit cooperation between riparian states over water resources? This is linked to the question of how institutions contribute to mitigate competition for natural resources and how states interact in a multilateral arena. Overall, the book argues that cooperation in transboundary river basins is possible even where there are asymmetric power relations, challenging realist assumptions about competition and conflict over resources.

**Andrea Haefner** is an Adjunct Research Fellow at the Griffith Asia Institute in Brisbane, Australia, where she also received her PhD. She has previously worked with international and national organisations in Asia and with the German local government. Her research focuses on non-traditional security issues, focusing on environmental security and water governance.

## Earthscan Studies in Water Resource Management

*For more information and to view forthcoming titles in this series, please visit the Routledge website: http://www.routledge.com/books/series/ECWRM/*

**Negotiating for Water Resources**
Bridging transboundary river basins
*By Andrea Haefner*

**Water Regimes**
Beyond the public and private sector debate
*Edited by Dominique Lorrain and Franck Poupeau*

**Legal Frameworks for Transparency in Water Utilities Regulation**
A comparative perspective
*By Mohamad Mova Al'Afghani*

**Land and Hydropolitics in the Nile River Basin**
Challenges and new investments
*Edited by Emil Sandström, Anders Jägerskog and Terje Oestigaard*

**Water and Rural Communities**
Local politics, meaning and place
*By Lia Bryant, with Jodie George*

**Trans-jurisdictional Water Law and Governance**
*Edited by Janice Gray, Cameron Holley and Rosemary Rayfuse*

**The Mekong: A Socio-legal Approach to River Basin Development**
*By Ben Boer, Philip Hirsch, Fleur Johns, Ben Saul and Natalia Scurrah*

**Rules, Norms and NGO Advocacy Strategies**
Hydropower development on the Mekong River
*By Yumiko Yasuda*

# Negotiating for Water Resources

## Bridging Transboundary River Basins

**Andrea Haefner**

Routledge
Taylor & Francis Group
NEW YORK AND LONDON

First published 2016
by Routledge
711 Third Avenue, New York, NY 10017

and by Routledge
2 Park Square, Milton Park, Abingdon, Oxon, OX14 4RN

First issued in paperback 2018

*Routledge is an imprint of the Taylor & Francis Group, an informa business*



*British Library Cataloguing in Publication Data*
A catalogue record for this book is available from the British Library

*Library of Congress Cataloging in Publication Data*
A catalog record for this book has been requested

ISBN 13: 978-0-367-00025-7 (pbk)
ISBN 13: 978-1-138-66632-0 (hbk)

Typeset in Times New Roman
by Sunrise Setting Ltd, Brixham, UK

# Contents

# Illustrations

## Figures



## Maps



## Tables

# Acknowledgements

This book is based on my PhD entitled *Cooperation and Conflict in Transboundary River Basins: The Mekong, Danube and La Plata River Basins* and would not have been possible without the assistance of many individuals and institutional entities. First, I would like to acknowledge Griffith University for providing me with a generous postgraduate scholarship, without which this research would not have been possible. This also comprises additional funding throughout the duration of my candidature, including fieldwork and grants for travel to South America, Europe and Southeast Asia.

I would like to thank friends, family and colleagues who supported me throughout my PhD and writing the book. My utmost thanks goes to my principal supervisor Professor Andrew O'Neil who provided me with feedback, encouragement and support over the duration of my research journey. This book would not have come to completion without the administrative and moral support I received from the Griffith Asia Institute. My deepest thanks go to Kathy Bailey, Meegan Thorley and Natasha Vary, for encouragement and support. My thanks also go to my previous PhD running mates, Greta Nabbs-Keller, Cathy Maloney and Bounthanongsack Chanthalath (Bo) who shared this long journey with me and provided invaluable assistance and friendship. My appreciation also extends to current and past researchers at the Griffith Asia Institute and lecturers at the School of Government and International Relations and the Department of International Business and Asian Studies.

In particular, I would like to thank Professor Jason Sharman, Dr Christian Wirth, Dr Line Liss, Stephen McLaughlin and Dr Daniel Halverson, for offering advice, help and support, even when it was not in their area of service or expertise. I would also like to thank Dr Paul Howard, Dr Doug Smith and Dr Tapan Sarker for keeping me in a job over the course of the PhD project and while writing the book. Importantly, this project would have not been possible without the support of my family and my partner and for their love, belief and support in me over the inevitable ups and downs of intense research.

This research project, moreover, would not have been possible without the assistance and support of staff and representatives at the Mekong River Commission (MRC), the International Commission for the Protection of the Danube River (ICPDR) and the Comité Intergubernamental Coordinador de los Países de la Cuenca del Plata (CIC). In addition, many interviewees kindly shared their thoughts and time during my fieldwork trips.

Furthermore, parts of Chapter 3 appeared in "Regional Environmental Security: Cooperation and Challenges in the Mekong Subregion," *Global Change, Peace and Security* 25, no. 1 (2013): 27–42 and in "Non-Traditional Security – the Case of Water Security in the Mekong Subregion," *Pacific Geographies* 40, July/August (2013): 17–22. I am grateful to these publishers for permission to use this material here. Finally, it goes without saying that I am solely responsible for any errors or shortcomings of this book.

# Abbreviations

**ADB** Asian Development Bank
**AEGE** ASEAN Expert Group on the Environment
**AEM–MITI** ASEAN Economic Ministers–Ministry for International Trade and Industry
**AEWS** Accident Emergency Warning System
**AMBDC** ASEAN–Mekong Basin Development Cooperation
**ANDE** Administración Nacional de Electricidad
**ANTAQ** Agência Nacional de Transportes Aquaviários
**APEC** Asia–Pacific Economic Commission
**APT** ASEAN Plus Three
**ARF** Asian Regional Forum
**ASCC** ASEAN Socio-Cultural Community
**ASEAN** Association of Southeast Asian Nations
**ASOEN** ASEAN Senior Officials on the Environment
**BCI** Biodiversity Conservation Corridors Initiative
**CARU** Administrative Commission for the Uruguay River
**CDRI** Cambodia Development Resource Institute
**CEFIR** Centro de Formación en Integración Regional
**CEO** Chief Executive Officer
**CEP** Core Environment Program
**CIA** Central Intelligence Agency
**CIC** Comité Intergubernamental Coordinador de los Países de la Cuenca del Plata
**CIH** Intergovernmental Committee for the Paraguay–Parana Waterway
**CLVM** Cambodia, Laos, Vietnam and Myanmar
**CMU** Chiang Mai University
**CPWF** Program on Water and Food
**DC** Danube Commission
**EAA** European Anglers Alliance
**EAS** East Asia Summit
**EBU** European Barge Union
**EC** European Community
**ECAFE** Economic Commission for Asia and the Far East
**ECE** Economic Commission for Europe
**ECSC** European Coal and Steel Community
**EGAT** Electricity Generating Authority of Thailand

| | |
|---|---|
| **EIA** | Environmental Impact Assessment |
| **EPDRB** | Environmental Program for the Danube River Basin |
| **EU** | European Union |
| **EUR** | Euro |
| **EVNI** | Electricity of Vietnam International Joint Stocking Company |
| **FCDI** | Forum of Comprehensive Development of Indochina |
| **FLACSO** | Faculty of Latin American Studies in Social Science |
| **FONPLATA** | Treaty of the Financial Fund for the Development of the La Plata Basin |
| **FTAA** | Free Trade Area of the Americas |
| **GDP** | Gross Domestic Product |
| **GEF** | Global Environment Facility |
| **GIZ** | Deutsche Gesellschaft für Internationale Zusammenarbeit |
| **GMS** | Greater Mekong Subregion |
| **GNRBS** | Gabcikovo–Nagymaros River Barrage System |
| **GTAN** | High-Level Technical Group |
| **IAD** | International Association of Danube Research |
| **IAEA** | International Atomic Energy Agency |
| **ICJ** | International Court of Justice |
| **ICPDR** | International Commission for the Protection of the Danube River |
| **IDB** | Inter-American Development Bank |
| **IFAD** | International Fund for Agricultural Development |
| **IFC** | International Financial Corporation |
| **IHP** | International Hydropower Program |
| **IMC** | Interim Mekong Committee |
| **IUCN** | International Union for Conservation of Nature |
| **IWRM** | Integrated Water Resource Management |
| **JC** | Joint Committee |
| **JCCCN** | Joint Committee on Coordination of Commercial Navigation |
| **kWh** | Kilowatt Hour |
| **LS II** | Lower Sesan II |
| **MBDS** | Mekong Basin Disease Surveillance |
| **MC** | Mekong Committee |
| **MEE Net** | Mekong Energy and Ecology Network |
| **Mercosur** | Southern Cone Common Market (Spanish: Mercado Común Del Sur) |
| **MoU** | Memorandum of Understanding |
| **MRC** | Mekong River Commission |
| **NA** | National Assembly (Laos) |
| **NAFTA** | North American Free Trade Agreement |
| **NATO** | North Atlantic Treaty Organisation |
| **NGO** | Non-Governmental Organisation |
| **NGOF** | NGO Forum |
| **NHI** | National Heritage Institute |
| **NPU** | National Planning Unit |
| **OAS** | Organization of American States |
| **OCHA** | UN Office for the Co-ordination of Humanitarian Affairs |
| **OSP** | Office of the Secretariat in Phnom Penh |
| **OSV** | Office of the Secretariat in Vientiane |

| | |
|---|---|
| **PKK** | Kurdistan Workers' Party |
| **PNPCA** | Procedures for Notification, Prior Consultation and Agreement |
| **PPP** | Purchasing Power Parity |
| **QEC** | Emerald Triangle Quadripartite Economic Cooperation |
| **RBO** | River Basin Organisation |
| **RCC** | River Coalition of Cambodia |
| **RCSD** | Regional Centre for Social Science and Sustainable Development |
| **REC** | Regional Environmental Centre for Central and Eastern Europe |
| **SEA** | Strategic Environment Assessment |
| **TEN-T** | Trans-European Transport Network |
| **TERRA/FER** | Towards Ecological Recovery and Regional Alliance/ Foundation for Ecological Recovery |
| **TWINS** | Transboundary Waters Interaction NexuS |
| **UN** | United Nations |
| **UNDP** | United Nations Development Program |
| **UNECE** | United Nations Economic Commission for Europe |
| **UNEP** | United Nations Environmental Program |
| **UNESCAP** | United Nations Economic and Social Commission for Asia and the Pacific |
| **UNESCO** | United Nations Educational, Scientific and Cultural Organisation |
| **US** | United States |
| **USAID** | United States Agency for International Development |
| **USD** | United States Dollar |
| **VPOA** | Vientiane Plan of Action |
| **WFD** | Water Framework Directive |
| **WTO** | World Trade Organisation |
| **WWF** | World Wide Fund for Nature |
| **WWF–DCP** | World Wide Fund for Nature – Danube–Carpathian Program |

# 1 Introduction

Over 90 per cent of the world's population lives in countries that share a river basin with others. Freshwater resources are scarce and different nations, actors and users compete for limited sources in transboundary river basins; often conflicting with each other. Water is a resource with no substitute: it cannot be secured in sufficiently large quantities through long-distance trade deals; and, due to the interconnectivity of the hydrological system, the actions of one country in its water management have a direct bearing on the interests of neighbouring countries.[1] For instance, in the Mekong River Basin in Southeast Asia, current hydropower and navigation developments in certain countries impact on traditional sources of income such as fisheries, and rice production in others. These kinds of changes in water use have given rise to conflict between countries in that region and others, but have also led, in some cases, to greater cooperation. The past few decades have seen a number of new agreements about the sharing of river resources and cooperation between riparian states.

This book explores the drivers of conflict and cooperation between states in transnational river basins. Drawing on extensive fieldwork and interviews on the Mekong in Southeast Asia, the Danube in Central and Eastern Europe and the La Plata River Basins in South America, the book provides a three-level analysis across three case studies, including the regional framework (European Union (EU), Association of Southeast Asian Nations (ASEAN) and the Southern Cone Common Market (Mercosur), the River Basin Organisations (RBOs) and the micro level. Overall, this research will test the commonly held assumption that when the upstream country is also the basin's hegemon there is little incentive to cooperate. Based on this, the key research question of the book is: To what extent do power asymmetries prevent or inhibit cooperation between riparian states over water resources? This is linked to the question of how institutions contribute to mitigate competition for natural resources and how states interact in a multilateral arena.

To do so, this book investigates the phenomenon of cooperation and conflict over international water resources. In particular, it focuses on transboundary river cooperation between riparian states with asymmetrical power relations, where the dominant power is an upstream country. Too often in international relations excessive focus is given to military and economic power, and there is an insufficient understanding of what constitutes other types of power.[2] As argued by realist scholars, cooperation in international relations is limited. This is especially the case in regards to water resource management, when the upstream country is the regional dominant power, as the asymmetric power situation is believed to make the whole bargaining game between countries more difficult. There exist two different scenarios for hydropolitics in asymmetric power relations. First, when the upstream country is also the basin's hegemon there is little incentive to cooperate. Second, when the downstream state is the hegemon, cooperation is possible; however, it depends on the hegemon and is controlled by it.[3] This study will focus on the first scenario and challenge this assumption.

Derived from the key research question and embedded in the existing literature, which is presented in Chapter 2, this book tests the following four hypotheses. The first hypothesis is that conflict is less likely if multilateral institutions include all riparian states and power distribution is equal, or favours the downstream country. The second hypothesis is that conflict is less likely if the existing multinational institutions have strong legal frameworks, including dispute resolutions mechanisms. The third hypothesis is that cooperation is more likely if a sense of regionalism exists. The fourth hypothesis is that cooperation is more likely if strong linkages between the states exist.

To test these hypotheses and to answer the main research question, the research provides a three-level analysis across three case studies (see Table 1.1). The three levels of analysis include the regional framework, the RBOs and the micro level, focusing on the selected case studies.

Based on the findings from the hypotheses, this book shows that realist explanations of power, defined in terms of military and economic capabilities, are not sufficient to explain relations in transboundary river basins. It shows that even when the upstream country is also the basin's hegemon, countries in river basins find reasons to cooperate, make agreements and create institutions to manage water resources. Taking into consideration the theoretical concepts of realism and liberalism, this research shows that besides the power relation between the riparians, issue linkage and bargaining plays a central role. Also of great significance is the influence of the regional umbrella – including historical ties, institutional design and political system. This is also linked to the importance of the regional level, which is often underemphasised in international relations, especially when focusing on power theories. Overall, the book argues that cooperation in transboundary river basins is possible even where there are asymmetric power relations, challenging realist assumptions about competition and conflict over resources. Further, this book argues that there are nuances between cooperation and conflict, and that this dynamic can change over time. For example, a river basin can be cooperative in some areas and less in other areas, as demonstrated by the empirical work of this book which shows that cooperation and conflict can exist in parallel.

*Table 1.1* Research framework

| *Case Study* | *1. Level of Analysis* | *2. Level of Analysis* | *3. Level of Analysis* |
|---|---|---|---|
| | *Regional Framework* | *Institutions* | *Micro Level (Within Case Studies)* |
| Mekong River | ASEAN | MRC | • Xayaburi Dam<br>• Lower Sesan II Dam<br>• Navigation on Upper Mekong |
| Danube River | EU | ICPDR | • Gabcikovo–Nagymaros Barrage System<br>• Baia Mare incident<br>• Bystroe Canal |
| La Plata River Basin | Mercosur | CIC | • Itaipu Dam<br>• Hidrovia Waterway<br>• Pulp Mill incident |

Notes: ASEAN: Association of Southeast Asian Nations; MRC: Mekong River Commission; EU: European Union; ICPDR: International Commission for the Protection of the Danube River; Mercosur: Southern Cone Common Market; CIC: Comité Intergubernamental Coordinador de los Países de la Cuenca del Plata.

In examining the four hypotheses, the research focuses on three case studies: the Mekong, Danube and La Plata river basins, which all have water-related security challenges, a dominant power as an upstream country and an existing institution dealing with water resource management. This book also includes a comparative aspect, questioning if the results vary depending on the regional context and the structure of the institution. Overall, natural resource management, or, more specifically, water resource management, was chosen as an example given that transnational challenges are of increasing importance. Moreover, transboundary rivers break territorial sovereignty and challenge policymakers and scholars. Because of this, the findings in this research can be used to draw broader conclusions for other transnational challenges in a multilateral environment, such as haze pollution or the spread of diseases.

## Significance of the Research

This research is important in two regards: first, the significance of the problem (i.e. the importance of water) and second, the contribution to the literature on the links between power relations and cooperation in transnational river basins. Empirically, this study is important because over 90 per cent of the world's population live in countries sharing a river basin. Water is a vital source that deserves special attention because of three key characteristics: it is a resource with no substitute; it cannot be secured in sufficiently large quantities through long-distance trade deals; and, due to the interconnectivity of the hydrological system, the actions of one country in its water management have a direct bearing on the interests of neighbouring countries.[4]

Previous studies in the book's area have ranged in focus from the analysis of one river basin to the analysis of several transnational rivers, including the ten largest rivers in the world. However, there are only a few existing studies that focus on a select number of transnational rivers and use a comparison across cases from different regions (transregional) to provide a detailed analysis of asymmetric power relations. None of them considers the three selected cases of this study. As such, this study builds on existing work but goes beyond it by providing important findings in the context of the wider literature on transboundary river basins, as this study combines theoretical knowledge with a transregional approach to provide a detailed analysis of a few in-depth case studies. Further, the study will add to the academic literature on cooperation regarding non-traditional security challenges, and potentially assist policymakers whose concerns focus on decision-making in a subregion. Additionally, the book shows how cooperation in one area can contribute to cooperation in other areas through spill-over effects. Overall, water resource management and allocation plays a crucial part in potential conflict between riparian states in most, if not all, regions in the world. As a result, the future developments of many countries depend on the successful management and allocation of these transnational resources, and particularly on the successful cooperation between riparian states regarding transboundary water management.

Theoretically, this study fills an important gap. The study adds to the academic literature on cooperation regarding non-traditional security challenges and it will be important for policymakers whose concerns focus on decision-making in a subregion and how cooperation in one area can contribute to cooperation in other areas. The research further unpacks the contemporary scholarly literature and contributes to ongoing debates about cooperation on non-traditional security issues, especially in regards to asymmetric environments. This will include existing literature on regional developments in general, but a more detailed focus will engage with the literature on the case studies and its implications for other regional developments on river cooperation. Overall, the research provides insight into why it is important to focus on different levels of analysis and link them together. It also illustrates how to achieve robust institutions that can adapt to changing environments and challenges that arise from increasing transnational security threats.

## Research Strategy

There are many different ways to approach the topic of cooperation and conflict in transboundary river basins. Civil society, non-state actors and a variety of political actors in different political environments are of growing interest; however, the main focus of this research is on riparian states. Similarly, a quantitative approach could have been employed to measure different variables; for example, based on the Database for Transboundary Freshwater Dispute. However, mindful that a quantitative approach would not lend itself to a three-level analysis of complex situations across three case studies, I have chosen a qualitative approach to investigate the research question through the selected case studies.

My research employs a mix of qualitative methods and includes information gathered from semi-structured interviews with experts on the river basins of the three selected case studies. The research further unpacks the scholarly literature and contributes to ongoing debates in regards to cooperation on transnational security studies, especially in regards to asymmetric power relations in transnational river environments. As well as the existing literature on regional developments, there is a detailed focus on the literature relevant to the case studies and its implications for other regional developments on river cooperation. Integral to this research are the semi-structured interviews, which provide a useful middle ground between unstructured and structured interviews.

Interview participants were drawn from a range of experts from the RBOs, the riparian governments, international organisations, academia and the non-governmental sector in the relevant countries. Interviews with local experts from each of the three river basins were held and included around a total number of 49 individuals interviewed.[5] Interviews were held from March 2012 to May 2014, and locations varied. In regards to the Mekong River, most interviews were completed in the capital cities; namely Vientiane in Laos; Phnom Penh in Cambodia; and Bangkok in Thailand. Interviews on the Danube River were conducted in Vienna, Austria; and Budapest, Hungary. For the La Plata River Basin, the main locations included Buenos Aires, Argentina; Montevideo, Uruguay; and Foz do Iguacu, Brazil.

Supplementary Skype and telephone interviews were conducted with other experts on transnational river basins located outside the river basins and also as follow-ups to the face-to-face interviews. Additionally, documentary studies were conducted, in particular through the analysis of publicly available government texts and reports; for instance, through the use of the extended library of the Mekong River Commission (MRC) in Vientiane, Laos and the library of the Cambodian Development Research Institute in Phnom Penh, Cambodia. This also included the content study of official documents, publications, media texts and other papers from the broader epistemic community in the three regions, but also from outside the regions. Analysis of foreign material was mainly in English. However, German texts and documents were also used for the Danube River case study and, importantly, Spanish literature on the La Plata River, as this is still the most used language regarding transnational river development in South America (besides Portuguese). Where I have included foreign-language quotes or references, an English-language translation has been provided.

## Case Study Selection – The Mekong, Danube and La Plata River Basins

Besides using interviews, the book uses a case study method to provide an examination of the selected single cases, but also includes a comparative aspect. Researchers use the case study method to examine selected single cases, but also to include a comparative aspect. This method is different to the comparative method, which only uses a comparison among a small number of cases.[6] For Jason Seawright and John Gerring, a case study is 'the intensive (qualitative or quantitative) analysis of a single unit or a small number of units (the cases), where the researcher's goal is to understand a larger class of similar units (a population of cases)'.[7] Further, case studies are useful because they are based on specific events and inform how occurrences and trends are unfolding.[8] Case studies are helpful when attempting to highlight particular decisions, processes, institutions and events.

For this to be valid, it is important that the cases are selected wisely and are able to be applied to other populations with similar characteristics.[9] Seawright and Gerring argue that 'most similar' cases that widely represent the population should include a minimum of two cases and will provide the strongest basis for generalisation.[10] More specifically, in regards to transboundary rivers, Shlomi and Ariel Dinar argue that a single case study is *ad hoc* and not sufficient to understand the importance of cooperation in transnational river basins, and that a broader focus is beneficial.[11] Rather than explicitly comparing the three river basins, this study provides an in-depth analysis of the case studies that illustrates the processes and negotiations involved in cooperation and how this impacts wider regional integration. The case study approach is appropriate for this research to tackle the questions of 'how' and 'why'; when subjects and/or events are not controlled by the researcher; and when researchers focus on contemporary events in their natural context.[12] These count especially for the micro-level analysis as they provide concrete practical examples.

This research will use 'most similar' cases, which means that the selected cases have similar specified variables according to the following criteria. All three cases have water-related security challenges, a dominant power (in regards to their relative economic and military power) as an upstream country and an institution dealing with water resource management. Further, the three case studies are all rivers over 2,780 kilometres in length, ranking in the top 20 rivers worldwide and, together, comprise 30 countries. The case studies are drawn from different regions and are used to draw wider conclusions on river basins worldwide with similar asymmetric power relations. In addition to the comparative case studies, the book also draws strongly on historical narrative to provide a more solid empirical basis for comparing changes over time.

Overall, transboundary water relations are complex and all basins are different because water resources' endowments, terrain and – most importantly – institutional infrastructure vary. While the nature of power relations is unique to each river basin,[13] in arid or semi-arid regions such as Africa and the Middle East, all the above-mentioned factors are intensified.[14] Additionally, due to the conflict-prone nature of these regions, water resource management is not high on the political agenda. Nevertheless, numerous studies focus on water resource management in the Middle East and Africa, including works on the following transboundary rivers: the Euphrates and Tigris,[15] the Jordan,[16] the Nile[17] and the Senegal.[18] Therefore, my study excludes these two regions.

Similarly, the measurement of a basin surface, river flow or river length is difficult to ascertain and depends on different measurement methods and whether or not river tributaries are included. As argued by Del Castillo, 'major basins and major rivers do not always coincide, but combining major river basins – those with more than three million square kilometres, and major rivers – which are more than 3,000 kilometres long, the greatest fluvial systems of the world can be identified' (Table 1.2).[19] However, due to the transnational aspect of this study, national rivers are not included and, as a result, the La Plata, Mekong and Danube rivers were selected.

The La Plata River Basin is the largest river system in South America and the fourth largest in the world, covering large parts of northern Argentina, south-eastern Bolivia and the southern part of Brazil, the whole of Paraguay and a large part of Uruguay. The basin has around 100 million inhabitants, 75 large dams and the river is core to the region's socio-economic activities, generating around 70 per cent of per capita Gross Domestic Product (GDP) of the five riparians.[20] Similarly, the Mekong River is the largest river in Southeast Asia and is ranked in the top ten longest rivers worldwide. The Mekong Subregion is an important area in the Asia–Pacific due to its significance for more than 70 million people who live directly on the river banks and the economic development of the six riparian countries – China, Myanmar, Laos, Thailand, Vietnam and Cambodia.[21] Unlike the Mekong and the La Plata, the Danube River is not among the ten largest or longest rivers worldwide, but was chosen because its historical and institutional developments are important for this research.

Moreover, the Danube River is the second largest river in Europe that provides an insight into transnational river basins. Additionally, the Danube River comprises a total area of 801,463 square kilometres, with more than 80 million people living on its banks and is the world's most international river basin, shared by 19 countries.[22] All of these three selected rivers flow through national borders of riparian states and their bio-geophysical nature requires coordination between the nation states in order to prevent conflict and achieve positive outcomes for the whole river basin. All three rivers are counted as large river basins and are ranked in the top 20 largest rivers worldwide.

This book includes the three above-introduced case studies which are each divided into three interconnected parts. The first part investigates the first level of analysis, the regional framework and its dynamics. It will focus on the relations between the riparian states and the interactions and interlinkages based on the regional framework. This is then linked to the second level of analysis, the RBOs,

*Table 1.2* Overview of large river basins

| ***Largest River Basins*** | |
|---|---|
| ***Basin Catchment Area*** | ***Million/km²*** |
| **Amazon** | 6.2 |
| **Congo** | 3.6 |
| **Mississippi/Missouri** | 3.2 |
| **La Plata** | 3.1 |
| **Nile** | 3.0 |
| **Ob** | 2.7 |
| **Yenisey** | 2.49 |
| **Lena** | 2.42 |
| **Lake Chad** | 2.3 |
| **Niger** | 2.1 |
| ***Rivers with the Greatest Flow*** | |
| ***River's Flow*** | ***m³/s*** |
| **Amazon** | 209,000 |
| **Congo** | 46,200 |
| **Ganges** | 43,800 |
| **Yangtse** | 31,350 |
| **Orinoco** | 28,700 |
| **La Plata** | 22,800 |
| **Yenisey** | 19,990 |
| **Mississippi** | 18,000 |
| **Lena** | 18,000 |
| **Mekong** | 15,770 |
| ***Rivers with Length of More Than 4,000 km*** | |
| ***River's Length*** | ***km*** |
| **Nile** | 6,695 |
| **Amazon** | 6,516 |
| **Yangtze** | 6,300 |
| **Mississippi-Missouri** | 6,275 |
| **Yellow** | 5,464 |
| **Congo** | 4,667 |
| **Amur** | 4,440 |
| **Lena** | 4,270 |
| **Mekong** | 4,350 |
| **Mackenzie** | 4,241 |
| **Niger** | 4,200 |
| **Yenisey** | 4,092 |

Source: adjusted from Del Castillo (2011).

focusing on the effectiveness of the institution and its design structure. The last part of the case studies investigates the micro-level analysis using key decisions made within the river basins. This is important as here power is measured as a result of key decisions made within a basin and their impact on the existing institutions and frameworks. In this context, the importance of issue linkage and bargaining are highlighted, as water and natural resource management cannot be seen as separate from other pressing issues between actors and riparians in a region.

## Chapter Outline

This book addresses the research problem in seven chapters. Following the introduction in Chapter 1, Chapter 2 expands on the introduction and provides a literature review and contribution to the field. Key analytical concepts discussed include cooperation and conflict in transboundary river basins, the role of power including bargaining and issue linkage, and the role of institutions before looking into the global and regional levels. It further details the research design of the book, involving a structured and focused comparison across time and regions.

Chapters 3, 4 and 5 include the empirical work and focus on the three case studies. The case studies are largely structured into three constituent parts to provide a comparative context: the first level of analysis, the regional framework; the second level of analysis, the RBOs; and the third level of analysis, the selected micro-level case studies. This structure provides the basis to understand the three levels of analysis and how they are interlinked and inform each other.

Chapter 3 introduces the first case study, the Mekong River. The first section highlights briefly the importance of the Mekong subregion and examines the historical development on the Mekong River over time. It reveals that it is important to understand the historical developments and political ties in the region in order to understand the whole progress of the subregion, including challenges such as wariness between the riparians based on past ideological divides. This section also investigates the role of the riparian countries in relation to the upstream–downstream location and their status within the region (including conflicting demands on limited natural resources). The second part of the chapter examines the multilateral and bilateral institutions in the Mekong region and places emphasis on the regional framework of the ASEAN before introducing the MRC and the Greater Mekong Subregion (GMS). The third part focuses on the micro-level analysis. The chapter uses the Xayaburi Dam, the Lower Sesan II Dam and the navigation in the Upper Mekong as case studies to show how issues are interlinked and influence decision-making in the Mekong subregion.

Overall, Chapter 3 shows that the regional umbrella plays an important role in regards to regional development and cooperation; however, this is limited in regards to natural resource management. The chapter further demonstrates that although the MRC is successful in data collection and research on water resources management, its mandate is limited by a very restricted mandate and recent processes including the Procedures for Notification, Prior Consultation and Agreement (PNPCA), which make it difficult for the MRC to successfully get involved in dam developments such as the Xayaburi Dam. Additionally, the GMS plays an increasingly prominent role and can be a driver for future successful cooperation. Further, it is important to understand the key actions and dynamics of this interlinked region in order to get a deeper understanding of the interconnectedness of decision-making and its results. Decision-makers and politicians involved, and their linkages, are often the drivers behind decision-making regarding large infrastructure developments. In each country this results in large differences between the middle classes and decision-makers on one hand, and the majority of the population on the other. Often decisions are not made because they are the best solution for the people, but for the benefit of the elites. Furthermore, there is still a very strong feeling of sovereign rights in all riparian states that adds to the difficulty of decision-making.

The second case study, the Danube River Basin, is introduced in Chapter 4. The chapter provides an overview of the historical background of the Danube region and the early river developments, as well as momentous major regional events during the Cold War before highlighting the role of the EU. It focuses especially on the integration and enlargement processes in 2004 and 2007, and the role of the aspiring riparian states with ambitions for EU membership in the near future. It also emphasises the role of EU environmental legislation, including the important Water Framework Directive (WFD), which, the chapter argues, has a strong influence on the Danube River and the riparian countries. Similarly, power relations in the river basin are closely linked to the industrial development of the riparian countries and historical context, as can be seen between Germany and Austria on the upstream of the Danube in comparison to former Yugoslavia and Romania on the downstream. The chapter further highlights the different important multilateral agreements and main institutions, their influence in the region and their handling of environmental challenges over the last decades, with a special focus on the International Commission for the Protection of the Danube River (ICPDR). The chapter then uses the Gabcikovo–Nagymaros River Barrage System, the Baia Mare cyanide spill and the Bystroe navigational canal as case studies to provide an insight into decision-making on three key decisions made in recent times and the role of the ICPDR in their regard.

Overall, Chapter 4 argues that although on the surface the Danube case appears very different to the Mekong and La Plata cases, the challenges are very similar, as the three micro-level case studies demonstrate. Geographical positioning has a strong impact on cooperation and the institutions along the Danube, and without a strong commitment to cooperation by both upstream countries the situation of the ICPDR would be very different. Additionally, the chapter shows that the ICPDR plays an active role focusing on mediation, also highlighting the role of public participation and non-state actors. The chapter reinforces findings that the EU framework plays a major role in the third level of analysis and has advantaged environmental standards and developments; however, there are still gaps which need to be filled, including the problematic adjustment of national legislation to international standards which can often take several years. Further, I argue that although cooperation on the Danube River is well developed and public participation exists, these benefits should be developed and increased further. In the Eastern European countries in particular there are still enormous economic, social, environmental and political gaps between riparian countries. European enlargement and integration play a crucial part in the developments along the Danube River and are key ingredients for further successful cooperation.

The final case study, presented in Chapter 5, investigates the La Plata River Basin. The chapter highlights the political, economic as well as ecological importance of the La Plata region for South America and the significance of future regional development for the wider region in order to reduce environmental degradation. As a result of its geographical location, developments and cooperation are closely linked to the past and current Brazilian–Argentina relationship, existing diverse treaties and regional developments including Mercosur. Before highlighting the different existing cooperative agreements, including the La Plata River Treaty as a key framework, the chapter focuses on the stakes for the riparian countries. Further, the chapter uses three case studies – the Itaipu hydropower dam, the Hidrovia waterway transportation system and the pulp mill incident on the Uruguay River – to provide an insight into decision-making.

Chapter 5 argues that although the La Plata River Basin exhibits some unique characteristics, such as several bilateral and multilateral agreements, the challenges are similar to the Mekong and Danube River cases, including increased hydropower developments and the importance of navigation, but also the challenges of pollution and flooding. Overall, this chapter argues that cooperation in regards to water resource development is based on different bilateral and multilateral treaties, which were established during two dominant phases in the 1970s and 1990s focusing on economic development on the La Plata River Basin. Problems include the less prominent role of public participation, underfinanced secretariats and a lack of communication and coordination between institutions to achieve a basin-wide approach. The chapter also shows that when it comes to economics and urgent issues, the bilateral way is the preferred option in the La Plata Basin. It also shows that natural resource management and river cooperation are used as political instruments and demonstrations of power seen, for instance, in the possible 'random' influence of Argentina on navigation and its impact on the other riparian states. Similarly, the pulp mill incident illustrates the role of Mercosur as a regional institution focusing on trade and free flow of goods, which, however, could not resolve the dispute and lacked the enforcement powers to re-open the international bridge. Overall, the chapter highlights that upstream–downstream relations and reliance on the river are crucial to understanding the bigger picture and comparing it to the other two cases.

Chapter 6 combines the framework chapter with the empirical work and provides a cross-case analysis. It begins by revisiting the dilemma and importance of water resource management before it examines the three levels of analysis, the regional framework, the RBOs and the specifics within the case studies. The brief conclusion in Chapter 7 includes a summary of this book and its key findings, and revisits the main hypotheses and the significance of this research. The book concludes with possible directions for further research and points out significant implications for policymakers and scholars.

## Notes

1 Loh Su Hsing, "China Holds the Key to Asia's 'Blue Gold'," *Jakarta Globe*, 15.12.2011.
2 Shlomi Dinar, "Power Asymmetry and Negotiations in International River Basins," *International Negotiation* 14, no. 2 (2009a): 329.
3 Miriam R. Lowi, *Water and Power: The Politics of a Scarce Resource in the Jordan River Basin* (Cambridge: Cambridge University Press, 1993); Shlomi Dinar, "Scarcity and Cooperation Along International Rivers," *Global Environmental Politics* 9, no. 1 (2009b): 110.
4 Loh Su Hsing, "China Holds the Key to Asia's 'Blue Gold'."
5 For an overview of the interviews see the Appendix – List of Interviews and Special Libraries.
6 Alexander George and Andrew Bennett, "Case Studies and Theory Development," in *Case Studies and Theory Development in the Social Sciences*, eds. Alexander George and Andrew Bennett (Cambridge, MA: Harvard University, 2005), 18–19.
7 Jason Seawright and John Gerring, "Case Selection Techniques in Case Study Research," *Political Research Quarterly* 61, no. 2 (2008): 296.
8 Jeffrey Stark, "Environmental Security in Latin America and the Caribbean: Development, Conflict, and Democracy," in *New and Critical Security and Regionalism – Beyond the Nation State*, eds. James J. Hentz and Morten Boas (Hampshire: Ashgate Publishing Limited, 2003), 71.
9 Seawright and Gerring, "Case Selection Techniques in Case Study Research."
10 Ibid.
11 Shlomi Dinar and Ariel Dinar, "Recent Developments in the Literature on Conflict and Cooperation in International Shared Water," *Natural Resources Journal* 43, no. 4 (2003): 1285–6.
12 Robert K. Yin, *Case Study Research: Design and Methods* (Thousand Oaks, CA: SAGE, 1989); Joe R. Feagin, Anthony M. Orum, and Gideon Sjoberg, *A Case for the Case Study* (Chapel Hill, NC: University of North Carolina Press, 1991).
13 Mark Zeitoun and J. A. Allan, "Applying Hegemony and Power Theory to Transboundary Water Analysis," *Water Policy* 10, no. 2 (2008): 11.
14 Marwa Daoundy, "Asymmetric Power: Negotiating Water in the Euphrates and Tigris," *International Negotiation* 14, no. 2 (2009): 359.
15 Ibid.; Frederick W. Frey, "The Political Context of Conflict and Co-Operation over International River Basins," *Water International* 18, no. 1 (1993): 54–68.
16 Thomas F. Homer-Dixon, "Environmental Scarcities and Violent Conflict: Evidence from Cases," *International Security* 19, no. 1 (1994): 5–40.
17 Ana Elisa Cascao, "Ethiopia – Challenges to Egyptian Hegemony in the Nile Basin," *Water Policy* 10, no. 2 (2008): 13–28; Peter H. Gleick, "Water and Conflict: Fresh Water Resources and International Security," *International Security* 18, no. 1 (1993): 79–112; Dinar, "Scarcity and Cooperation Along International Rivers."
18 Homer-Dixon, "Environmental Scarcities and Violent Conflict: Evidence from Cases."
19 Lilian Del Castillo, "The La Plata Basin System against the Background of Other Basin Organizations," *International Journal of Water Resources Development* 27, no. 3 (2011): 515–16.
20 Victor Pochat, "International Agreements, Institutions and Projects in La Plata River Basin," *International Journal of Water Resources Development* 27, no. 3 (2011): 501.
21 Katri Mehtonen, Marko Keskinen, and Olli Varis, "The Mekong: IWRM and Institutions," in *Management of Transboundary Rivers and Lakes*, eds. Olli Varis, Asti K.

Biswas, and Cecilia Tortajada (Berlin: Springer, 2008), 210; Central Intelligence Agency (CIA), "Laos," www.cia.gov/library/publications/the-world-factbook/geos/la.html (accessed 23.11.2015); CIA, "Cambodia," www.cia.gov/library/publications/the-world-factbook/geos/cb.html (accessed 23.11.2015).

22 International Commission for the Protection of the Danube River (ICPDR), "The Danube River Basin – Facts and Figures," ICPDR, Vienna, 2009, 7.

# 2 Cooperation and Conflict in Transboundary River Basins: A Framework

Conflict between and within states over freshwater resources poses an interesting challenge for scholars and policymakers due to competing claims over water and the concept of territorial sovereignty that is broken by transnational rivers. There are 261 transboundary river and lake basins worldwide, which cover nearly half of the earth's land surface.[1] This accounts for approximately 60 per cent of the global freshwater flow, and over 90 per cent of the world's population lives in a country that shares a river basin.[2] Most of the world's shared water resources remain outside transboundary agreements between all riparian countries, and some have no agreements in place.

While freshwater resources are in theory renewable, in reality they are finite, poorly distributed and often controlled by one state or group. Additionally, while freshwater is fundamental for all ecological and societal activities, such as energy and food production, industrial development, transportation, health and employment, it is often unequally distributed within a region or even within one country.[3] In China, for example, south and south-eastern parts of the country have sufficient water, while water scarcity is a major problem in the northern and eastern parts. As a result, the future developments of many countries depend on the successful management and allocation of these natural resources, including successful cooperation between riparian states in regards to transnational water management.

Some key problems related to water management include water shortages, inefficient water use, inadequate sanitation, destruction of the ecosystem, water logging and salinisation, and water pollution or flooding of cultivated, urban or industrial areas. Not all of these major problems are relevant in all transboundary river basins; challenges vary. For instance, in the past, the Danube, Rhine and La Plata rivers experienced major problems with water pollution, whereas in the Euphrates River water quantity remains a key challenge between the riparian states. On the Mekong River, water governance is closely related to environmental effects, especially regarding water quality and water flow. For instance, damming on the upstream river not only reduces the water flow influencing the problem of water scarcity or allocation in the downstream countries, but also influences the water and soil quality. This is interlinked with growing energy demands, which is one of the other major aspects of transboundary river basins. Problems arise when hydropower developments conflict with environmental issues.

Furthermore, food and energy depend on access to water, which adds to the pressures facing transnational river basins. This can be seen in the proposed controversial outline of the Hidrcvia Paraguay–Parana project in the La Plata River, a massive and complex navigation system which would allow year-round shipping of barge trains, but would also have severe implications for the world's largest wetlands. Overall, water resource management and allocation plays a crucial part in potential conflict between riparian states in most, if not all, regions in the world. This is why it is important to use a transregional approach in this study.

For this research, 'transboundary rivers' mean rivers that include more than two riparian states, based on the definition of 'transboundary waters' in the United Nations (UN) Water Convention, meaning 'any surface or ground waters which mark, cross or are located on boundaries between two or more States'.[4] Hence, any problems that arise are often within a multilateral context as opposed to a domestic or bilateral one. This does not include rivers that cross intra-national borders, such as the Murray Darling River Basin in Australia, which crosses through several states. Further, this definition is different to a 'watercourse', which often does not include the whole extent of a basin and its water sources, focusing more narrowly just on the water flow. Some scholars use 'international river' to refer to freshwater basins within the border of more than one state. However, this can be misleading since, for instance, in the concept of international high sea everyone has a right. In an international river, however, only riparian countries have a right. Additionally, an 'international river' is often linked back to the navigation on a river, which can be related to the Peace Treaty of Versailles that termed the Elbe, Oder and Danube rivers as international rivers.[5] Also, international rivers are rivers that form the boundary/boundaries between two or more nations but do not flow across them.[6] Similarly, the term 'shared waters' is also disputed as not all rivers are 'shared': conflicts may exist. As a result, the phrases 'transboundary rivers' and 'transnational rivers' are used in this book.

## Cooperation and Conflict in Transboundary River Basins

Natural resource management, and water management in particular, has received increasing attention from practitioners and scholars since the early 1990s. Arun Elhance defines this area of 'hydro-politics' as 'the systematic study of conflict and cooperation between states over water resources that transcend international borders'.[7] Initially, the research focus was on water wars and possible future conflicts over water resources between states in transboundary rivers. This has changed in recent years to a focus on cooperation in transboundary basins based on extensive quantitative and qualitative research led by Aaron Wolf, which shows that cooperation is prevailing in international river basins.[8]

Scholars focusing on conflict in transboundary river basins often refer to scenarios in the Middle East or Sub-Saharan Africa, where conditions are harsher and water is scarce. For instance, Peter Gleick uses the Jordan, Nile and Euphrates Rivers as cases for likely conflict due to growing competition for limited water resources and/or because of possible disputes over the ownership and rights to use the water.[9] Gleick further argues that disputes may evolve due to the contamination of shared water by upstream countries, for instance, on the Mekong or Rhine rivers; due to complications of managing multiple interests in a river, as in the Danube River; or because of difficulties sharing hydropower generation, as in the Parana or Zambezi river basins. Whether water scarcity or security can be seen as a national security threat as well as an environmental threat depends on several factors. According to Miriam Lowi, these include water dependency on one special river and the physical location of the country in relation to the river. Further factors are the countries' climate and condensation, the demands of a growing population and the political relations between the given country and the region.[10] Along similar lines, Frederick Frey argues that the three key elements leading to conflict include the importance of the water to the nation, relative military power and the riparian's position. Frey argues that the most conflict-prone situation occurs when the downstream riparian is most powerful and has the strongest interest in water, citing the Nile River and Egypt as an example.[11]

Expanding on this, Peter Gleick argues that the severity and urgency of a possible conflict or resolution depends on the following criteria: the degree of water scarcity experienced by the respective basin riparians; the extent to which the water supply is shared by more than one state; the ease of access to alternative freshwater sources (for instance, other major rivers); and the perceived over-exploitation and degradation of the water source by one riparian at the expense of another. To these he adds the protracted conflict that may underlie the water dispute; the historical and political claims made by each country over water; the relative power of the parties; and the geographic location of the riparians in relation to one another.[12] Practically, this means that China's role in the Mekong, as the most powerful country on the upstream with several other water resources, is different to the role of the most downstream country, Vietnam, which has a rapidly growing population and a strong focus on agriculture, being one of the largest exporters of rice.[13] Similarly, water scarcity could also develop into a national security threat in Cambodia and Paraguay, which depend 82 per cent and 70 per cent respectively on water originating from outside their borders.[14]

Besides scholars, politicians have also discussed transboundary water challenges. The former Israeli Premier Levi Eshkol declared in 1965 that 'water is a question of life in Israel' and 'Israel would act to ensure that the waters continue to flow'. Similarly, Herodotus states that 'Egypt is the Nile and the Nile is Egypt',[15] which demonstrates the importance of the river for the country. Although there has not been an international 'water war' yet, water can be a catalyst for a crisis as it was between Syria and Iraq in 1974.[16] However, in the

United States (US) century, only seven 'minor skirmishes' have occurred over water; in contrast, 145 water-related treaties were signed.[17] This does not mean that water wars do not exist, but conflict and turmoil over water is more often internal than international and occurs mainly between different tribes, or competing water users. This also implies that there is a connection between water and political stability and that a lack of clean freshwater for instance can lead to instability.[18] Furthermore, whereas violent conflict is rare, political disputes and conflict of interests over shared freshwater are not.

In contrast to water conflicts or water wars, some scholars see cooperation as the prominent contemporary development in transboundary rivers. Aaron Wolf presents four reasons against the plausibility of future wars, including arguments based on history, strategic interests, shared interests and institutional resilience.[19] Adding to this, Asti Biswas highlights the importance of trust/mistrust in regards to achieving cooperation between countries in South Asia. Biswas uses the Ganges-Brahmaputra–Meghna Basin as an example to highlight possible alternative outcomes. On the one hand, successful cooperation between India and Bhutan is leading to economic growth and a reduction in poverty. On the other hand, Biswas uses the example of India, Nepal and Bangladesh to show that mistrust can hinder cooperation and lead to the deprivation of a region through underdevelopment.[20] Biswas adds other interrelated factors that can be problematic in achieving agreements between riparians. These include political mistrust, historical rivalries, asymmetric power relations, increasing nationalism, short-term political party interests, growth of religious extremist groups and negotiations exclusively on water issues in the absence of a properly formulated negotiation framework. Moreover, the emergency of other issues between riparians and the influence of single-issue Non-Governmental Organisations (NGOs) can play a role.

Another group of scholars focuses on the role of institutional design in transboundary rivers and emphasises the structure, membership and function of River Basin Organisations (RBOs). Since the 1990s, there has been an increase in institutions at the river basin level, which helps to achieve cooperation between upstream and downstream countries, standardise national water policies and bring together all stakeholders with the goal of overcoming management issues.[21] Susanne Schmeier *et al.* focus on a theoretical definition and conceptualisation of RBOs and use three broad categories of constitutive elements, including internationalisation (binding international agreement over internationally shared river/lake); institutionalisation (permanence, infrastructure and actor quality); and governance (principles, norms, rules and mechanisms).[22]

Adding to this, other scholars focus on the formation of RBOs and their variations, and expand on the performance and effectiveness of governance in river basins. For instance, Ramiro Berardo and Andrea Gerlak focus on crafting a model for institutional effectiveness and study institutions that set the conditions under which river basins are jointly managed. The model focuses on two levels, the first one investigating the interstate agreement itself and the second addressing the process design elements, including transparency, scientific learning, conflict resolution and public input and representation.[23] Berardo and Gerlak use the conflict between Argentina and Uruguay to showcase how institutions can fail to be effective and how third-party intervention can be necessary to reach a solution. They link this to the failure of RBOs to adapt to changing circumstances and the need to focus more on the design attributes to achieve the legitimacy and effectiveness that would eventually lead to balanced management. This is also linked to Bernauer's work on explaining success and failure in international river management and the problematic of defining effectiveness.[24]

A third group of scholars, mainly based in the London School of Economics, focus on the concept of hydro-hegemony. This includes the role of hegemony, power and political economy processes in regards to transboundary water relations. This group of scholars argue that asymmetric power relations represent the key feature in the analysis of hydropower politics, including the hegemonic position and the uneven control of water resources between the riparians. They base this on the three dimensions of power established by Steven Lukes' *Power: A Radical View*, namely structural, bargaining and ideational power.[25] This also includes recent work by Naho Mirumachi focusing on transboundary water interactions by developing and applying the Transboundary Waters Interaction NexuS (TWINS) framework.[26]

Furthermore, Claudia Sadoff and David Grey use four types of cooperative benefits when examining the choices between conflict and cooperation: benefits to the river; benefits from the river; reduction of costs because of the river; and benefits beyond the river.[27] Sadoff and Grey further argue that a linkage between the different benefits from cooperation suggests that 'making a start in the environmental (Type 1) or direct economic cooperation (Type 2) can lead to growing political (Type 3) and indirect economic cooperation (Type 4) or vice versa'.[28] Adding to this, it is also important to remember that the topic of international river basins is not only driven by theories of international relations and politics but also several other disciplines, including studies such as an economic analysis of conflict, cooperation and institutions in international water management by Ines Dombrowsky, including theories of external effects, non-cooperative game theory and transaction costs economics.[29]

While cooperation and conflict are seen by some as opposing ends of the spectrum, I acknowledge that there are different levels of cooperation, which can range from notification processes and data exchanges to more sophisticated levels, including institutional designs. Cooperation means working together for common benefits, which implies coordination of behaviour among actors to realise at least some common goals.[30] Conflict can include, in extreme cases, war and armed conflict; however, it can also manifest in lesser forms over water allocation, hydropower projects or other disagreements. As stated by Mark Zeitoun and J. A. Allan, 'at all levels of hydro-politics, individual water users contend their interests and engage in conflictual relations'.[31]

My project concurs with Zeitoun and Mirumachi that 'cooperation and conflict actually co-exist, and perpetuate the paradigm that any conflict is "bad", and that all forms of cooperation are "good", meaning that there are different levels of conflict and varying levels of cooperation'.[32] Zeitoun and Mirumachi focus on 'water interactions' being defined as 'relations of co-existing cooperation and conflicts among communities, groups or states over international or sub-national waters, with a focus here on inter-state interactions'. This book is concerned with both areas – cooperation and conflict – and argues that there are nuances between cooperation and conflict in transboundary river basins today, and that these can change over time. For example, the fact that a river basin can be cooperative in some areas and less in other areas will be demonstrated further in the empirical work of this book, especially Chapters 3, 4 and 5.

### *The Role of Power in International Relations: A Realist Viewpoint*

The state-centred view of institutional order in the world goes back to Thomas Hobbes and Hans Morgenthau and persists in the post-1945 order in international relations. Hobbes was concerned about conflicts between groups within the states and drew the well-known conclusion in *Leviathan* that 'the natural condition of a man is a state of war'.[33] The power balance between the US and the Soviet Union during the Cold War reinforced this worldview, and the discipline of international relations has focused mainly on the relationship among states. According to Robert Keohane and Joseph Nye, three assumptions underpin the realist theory of international relations: states are coherent units and are dominant actors in international politics; force is a usable and effective instrument of policy in international politics; and issues in international politics are hierarchical, as in order of high politics and low politics.[34] Whereas during the Cold War superpower tension between the US and the Soviet Union was the key element in a bipolar world order, today regional security is increasing in importance. This idea is supported by Peter Gleick, who argues that the regional level plays an important role as conflicts are more likely to occur on local or regional levels in developing countries because common property resources may be more critical to survival and less easily replaced or supplemented.[35]

There is general agreement in the fields of political science and sociology that power is an important concept. Observers of political life agree that understanding the nature of power is vital to understanding political behaviour leading to decision-making. Hans Morgenthau provided a useful summary of the importance of power when he contended that 'international politics, like all politics, is a struggle for power. Whatever the ultimate aims of international politics, power is always the immediate aim'.[36] While there is not a universally accepted definition of power in the social sciences, it is often viewed as a relationship between two units.

Sociologist Max Weber offered a useful definition, describing power as 'the probability that one actor within a social relationship will be in a position to carry out his own will despite resistance, regardless of the basis upon which this probability rests'.[37] According to Robert Dahl, power is 'the ability to get others to do what they otherwise would not do', or, more generally, power is the ability to achieve one's purposes or goals.[38] In the relational power debate of the twentieth century, scholars have also tended to treat it as a causal variable. Dahl described the nature of power as 'A has power over B to the extent that he can get B to do something that B would not otherwise do'.[39] This view of power is accepted by the majority of authors who write about power relationships, even if their definitions are more descriptive.

For instance, Inis Claude describes power in the context of the balance-of-power approach and argues that balance of power is 'a system based on sovereignty and the absence of a world government, an anarchical structure'.[40] The realists' focus is on the interaction between states, especially the use of power and conflict between them. Most realists believe that international affairs are a struggle for power among self-interested states and individuals are always driven to pursue their own individual goals independently of their social context.[41] Adding to this, John Mearsheimer argues that in the twenty-first century states fear each other and seek to gain power at each other's expense because anarchy still exists after the Cold War.[42] He states that the 'fortunes of all states [. . .] are determined primarily by the decisions and actions of those with the greatest capabilities'.[43] Annette Freyberg-Inan further argues that the most important type of power resource is the state's military capability and the highest goal is to achieve its own security, which every state should maximise in relation to other powers.[44]

These assumptions can be linked to realist and neo-realists, including Robert Gilpin, Hans Morgenthau and Kenneth Waltz, who focus on the relative distribution of power as a function of capabilities. Relative power of the state is measured in military and economic capabilities. As a result, sources of structural power can be based on economic and military resources possessed by states, or, in the case of water resource management, in the hands of riparian states. Structural power is a material form of power, including military and economic power, and it can feature important variables such as major powers in the basin, distribution and patterns of military, economic and political powers, and regional and international alliances in trade, military and political affairs.[45] Marwa Daoundy develops this notion further by including the riparian position in the river basin and the controlled exercise over military, economic resources and water or 'the capacity to "turn the tap", as sources of structural power' in international river basins.[46] 'Realpolitik' argues that material resources are the defining factor in the power of a nation.[47]

Resulting from these power definitions, a hegemonic state can be defined as a state with a preponderance of power and a long-term view of interest, with both the capacities and incentives to create and manage a stable political order.[48] Hegemonic characteristics play an important role and can be distinctive in political institutions, culture and historic experience.[49] Miriam Lowi defines a weaker state thus: 'since the asymmetry of power is not in its favour, it is not in a position to achieve its aims and satisfy its need in an optimal fashion. Its capabilities are inferior to those of its adversary'.[50] Lowi argues this mainly in relation to the Euphrates–Tigris River where Turkey is the upstream state and also the dominant power. She further argues that, as a result, cooperation is least likely to materialise, as there is no incentive for Turkey to cooperate.[51] This results in the assumption that in most cases upstream hegemons are less likely to cooperate because they can utilise the water source unilaterally, resulting in an indefinite conflict of interest between the riparian states.[52] A second scenario is that when the downstream state is the hegemon, cooperation is possible; however, cooperation depends on the hegemon and is controlled by it.[53] John Mearsheimer argues in regards to asymmetric power relations that 'multipolar systems that contain especially powerful states [. . .] are the most dangerous system of all' and that the 'fortunes of all states [. . .] are determined primarily by the decisions and actions of those with the greatest capabilities'.[54] As stated earlier, Lowi argues that the riparians' geographical location along the river plays a crucial part in the power structure.[55]

Taking this knowledge of power to transboundary rivers, scholars in the London School of Economics focus on the concept of hydro-hegemony. The key features of the concept include the role of hegemony, power processes and political economy processes regarding transboundary water relations. They argue that asymmetric power relations represent the key feature in the analysis of hydropower politics, including the hegemonic position and the uneven control of water resources between the riparians. They base their conclusions on the three dimensions of power established by Steven Lukes, namely structural power, bargaining power and ideational power.[56] Ana Elisa Cascao develops this further by exploring the concept of counter-hydro-hegemony with a focus on the Nile River Basin, especially Ethiopia's role in it.[57] Her key argument is that hydro-hegemony is contestable and can change over time. Focusing on the case of Ethiopia, Cascao develops seven counter-strategies for a non-hegemon to achieve successful change in an environment with asymmetrical power relations: reactive diplomacy, active diplomacy, cooperation, mobilising international funding, building expertise-based knowledge, providing a discourse alternative and claims for legal principles.[58] This links to Frederick Frey, who argues that the analysis of power has different aspects and facets, including, for instance, the actors involved, their motivation, cognitions and structure (the nation's structural position within the world's political system).[59] It is also important to realise that power only becomes a potential resource if its control is seen to be valued by other actors in the interaction; meaning that power emerges from the relations between riparians, not from the power holder alone.[60]

### *Bargaining and Issue Linkage*

Bargaining is at the centre of many important issues in international politics and is often traced back to Thomas Schelling's *The Strategy of Conflict* and Ariel Rubinstein's work on the bargaining model.[61] Robert Powell developed this further; he examines multilateral bargaining and highlights the trade-offs that states face when deciding whether to align with the stronger power or with the weaker state, which puts the balancer in a stronger position relative to its coalition partner.[62] Making this political decision plays an important role in an asymmetric transnational river context, since the weaker downstream countries need to be strategic in their bargaining so as to ensure their needs are met adequately, whether in regards to a certain amount of river flow, an agreed level of water quality or other benefits from the river. Bargaining power, linked in part to material power, is an immaterial form of power that focuses on diplomatic and ideational capacities in order to impact on decisions and negotiation processes.[63] This power can be used by non-hegemonic powers to influence the hegemon. Marwa Daoundy includes the use of time, external alliances and the relative costs of no agreements as sources of power.[64] Bargaining power can also be related to the concept of 'soft power' developed by Joseph Nye.[65]

Shlomi Dinar focuses on bargaining power for weaker states in transnational environments through the realisation by both parties (mainly the powerful party) that efficiently harnessing the river requires coordination, linkage of other issues (water or non-water related) and the exploitation by the weaker parties upstream.[66] In a transboundary river, this can be the coordination of a hydropower cascade during the dry season or a linkage to other large infrastructure projects. This is linked to the definition of cooperation and conflict. Similarly, a separation must be made between the state level and the actors within it. Oran Young's concept of institutional bargaining is based on negotiations among self-interested parties to handle collective action problems in an international society. Young argues that such bargaining can only succeed when all the major parties and interest groups have a sense that their primary concern has been treated fairly.[67] Ernst Haas further develops the theory, arguing that collaboration only becomes a problem if the participating parties disagree on the distribution of the benefits: 'As long as the parties agree on the benefits to be derived from collaboration, conflict does not arise'.[68] This can be seen, for instance, in the detailed micro-level analysis of the pulp mill incident on the Uruguay River and its impact on the relationship between Argentina and Uruguay (see Chapter 5).

Another important aspect of bargaining power is 'issue linkage', meaning the linkage of other issues between states in relation to the situation over water resources. As a result of issue linkage, cooperation can be achieved through mutual concessions and/or the linkage into a 'bigger basket' with multiple benefits for the riparian countries involved. In this context, it is important to realise that cooperation and relations between states do not only focus on one issue; often, several areas are included, such as economic trade and investment, or security alliances. Thus, cooperation cannot be seen as isolated from states' interactions with one another, be it in bilateral or multilateral agreements. For instance, disputes between China and Vietnam in regards to the South China Sea can impact on both countries' relations on the Mekong.

In a broader context, environmental questions can also sometimes play an important role in trade, foreign aid and other international arrangements. For instance, in the North American Free Trade Agreement between Canada,

Mexico and the US (which came into force in 1994), environmental issues played a large role in the negotiations.[69] Issue linkage between environmental or water matters can include material incentives, such as financial aid, technology transfer to developing counties or trade sanctions, as seen in the Montreal protocol.[70] Expanding on this, Young argues that 'linking together disparate issues sometimes opens up possibilities for mutually acceptable arrangements by creating opportunities for the international equivalent of logrolling and the formation of package deals', which he highlights in regards to the incorporation of polar bears and trade in endangered species in the 1982 Law of the Sea Convention.[71] Haas defines issues as 'separate items that appear on the agenda of negotiations'.[72] Tactical linking includes issues that are not connected by any intellectual coherence to obtain additional bargaining leverage. For instance, on the Euphrates River, a security matter in regards to the Kurdistan Workers' Party (PKK)[73] movement in Turkey allowed Syria to achieve a bargain on the minimum amount of water flowing in the country from upstream dominant power Turkey. A possible further issue linkage between Turkey and Iraq can be seen in an exchange of oil for water. In regards to the Euphrates and Tigris rivers, 'power asymmetries do not necessarily determine the results of negotiations. But the impact of power is limited as negotiation outcomes are bilateral and temporary'.[74]

Overall, issue linkage can be seen as an effective bargaining power tool, especially for the weaker riparians. Similarly, a crucial prerequisite for resolving conflict is linking the problem to other aspects of the bilateral or multilateral relationship. For instance, if countries have a good bilateral relationship and cooperate on some issues, similar processes can be used in regards to water resource management. Haftendorn further argues that water conflicts need to be linked to other problems to replace an asymmetrical structure with a more cooperative symmetrical structure, including the possibility of trade-offs between economic, social and political benefits.[75] However, this is limited and only counts if the states have something to link that the other states value as important. Moreover, conflict settlement is eased if a high level of interdependence exists among the different parties of the conflict. This is linked to the general relations between the countries and possible existing tension in other areas. It is also linked to other existing multilateral institutions and to the existing regional framework. In general, the most common non-water linkages in transboundary river basins include financial compensation, land deals and political concessions.[76]

Besides issue linkage, payoffs also play a role in transnational river cooperation; for example, in the case of the Rhine River, the Netherlands paid compensation to France in the 1970s in order to achieve a better water quality flow, which allowed it to run its main port and also operate important national industries (such as the export of flowers). Based on these compensations, further negotiations increased international cooperation between the riparian states on the Rhine and resulted in the Rhine Chloride Convention 1976.[77] Daniel Deudney argues that resource scarcity caused by environmental degradation encourages joint cooperation to halt such outcomes, as degradation contributes to a network of common interests and goals; however, this depends on the interests of the riparians.[78]

As established earlier, this book argues that cooperation in transboundary river basins is possible in asymmetric power relations and that the statement that material power capabilities inevitably determine outcomes is limited. The key actors are the drivers behind achieving cooperation or conflict and are not necessarily the same as the riparian state as a whole. This means that, besides the asymmetrical relations between the riparians, the symmetrical relations within them influence regional institutions and actions. This point is further demonstrated by the importance of issue linkage and bargaining at all levels of the analysis. My research argues that power is important in international relations, but that material power is not the only form of power that can be used by states or actors within the state. Rather, bargaining power plays an important role in politics, usually in conjunction with material power resources which can include issue linkage. This view contrasts with many neo-realist and neoliberal theories that argue material power resources can be measured, and that state influence should be judged on the basis of military and economic resources.

### *Cooperation and the Role of Institutions: An Institutional View*

Based on the above realist argument about power, Stephen Krasner and Robert Gilpin argue that specific arrangements come into existence when those holding the power take the necessary steps to create them.[79] They further argue that the presence of a hegemon is necessary to establish institutional arrangements at the international level. Realists such as Joseph Grieco argue that cheating and the worry about relative gains are the major problems of institutional arrangements.[80] Kenneth Waltz further emphasises that 'the first concern of states is not to maximise power but to maintain their position in the system'.[81] Realists focus on the goal of attaining not the highest possible individual gain or payoff, but instead to prevent others from achieving advances in relative terms. Further, Waltz argues that states are preoccupied with power and security, and focus on conflict and competition rather than cooperation. Similarly, Hans Morgenthau argues that international institutions only slightly affect the prospects for cooperation.[82] As a result, international organisations are, according to realists such as Charles Glaser and John Mearsheimer, of little interest and simply reflect the national interest and power of the dominant states.[83] In regards to cooperation in power asymmetric environments, Mearsheimer argues that 'multipolar systems that contain especially powerful states [. . .] are the most dangerous system of all' and that the 'fortunes of all states [. . .] are determined primarily by the decisions and actions of those with the greatest capabilities'.[84] According to the realist view, cooperation is difficult to initiate or sustain, as its prospects are limited. Overall, however, cooperation is possible. Importantly, cooperation can emerge between states and is either the result of a collaborative arrangement with equitable gains and balanced for each of them, or is a reflection of the distribution of power between the parties.[85] One example of this ongoing divide between realists and liberals is the strong emphasis of realism on relative gains as an important consideration for states, whereas liberal theorists believe relative gains are less important than the absolute gains for each state in cooperation.

The liberal perspective, as distinct from the realist viewpoint, is relatively optimistic about the potential for increasing international cooperation and the future success of international regimes and other institutions. Liberal institutionalists believe that institutions can create peace out of international anarchy and that multilateral relationships are helpful in creating cooperation and interaction between conflicting states. They argue that institutions like the UN, the Association of Southeast Asian Nations (ASEAN) or the ASEAN Regional Forum assist in countering transnational security challenges such as terrorism, global warming and humanitarian crises, and manage them in a multilateral context. Moreover, domestic politics and the influence of actors within the states are important to foreign relations because state preferences are vital and dictate most of the government's international decisions.[86]

Furthermore, liberal theories of international relations focus on international institutions and economic interdependence for the support of cooperation and political stability. Strong trade and economic interdependence between regional states also play a major role in cooperation and the decrease of possible conflict, as argued by liberal theorists. For instance, Robert Keohane argues that regimes and international institutions foster cooperation because they reduce barriers, such as transaction costs or uncertainty that sometimes block mutual beneficial agreements between states, which in turn facilitate political stability.[87] Institutions also foster greater transparency, adding to the positive outcomes of cooperation.

Adding to this, Mark Levy *et al.* argue that cooperation 'requires effective international institutions to guide international behaviour toward sustainable development'.[88] Intergovernmental institutions can reduce power asymmetry between states by regulating reciprocity and cooperation.[89] Institutions mainly act because of three key problems: the low level of governmental concern about the environmental threat; lack of capacity to manage it; and the inability to overcome problems of collective action. These problems can be a driver for creating intergovernmental organisations, especially in regards to transnational environmental matters in which state borders do not stop the challenge.[90] In *After Hegemony* Keohane argues that 'cooperation under some conditions develop on the basis of complementary interests, and that institutions, broadly defined, affect the patterns of cooperation that emerge'.[91]

Overall, cooperation can include different forms and interests, but generally it can be defined as the 'coordination of behaviour among actors to realize at least some common goals'.[92] In this book, cooperation is based on this definition; more specifically, it means that on transnational rivers, cooperation can happen if all the riparians are in a RBO. However, cooperation can also occur if riparians cooperate outside of the institutions; for instance, in regards to data sharing or cooperation in specific areas only.

Moreover, the existence of a bilateral and multilateral treaty and its form impacts on the conditions on the ground. This is also the case if there are no treaties available. Some river basins have no basin treaty, some have sub-basin treaties and others have both basin-wide and several sub-basin treaties. For instance, the 15 treaties signed regarding the La Plata River Basin between 1946 and 2006 include three basin-wide agreements including all riparians,[93] and 12 agreements signed by only two or three riparian states. Similarly, the regional framework can vary significantly between regions, ranging from a strong legislative framework, as in the case of the European Union (EU) with binding agreements impacting on the river basins, to less formal ones, as is the case in Southeast Asia. In addition to power relations, Barry Buzan highlights the relevance of 'amity and enmity', meaning the importance of friendship, genuine support and expectation of support, and suspicion and fear. Buzan includes historical and geographical roots and, to a certain degree, cultural and racial ties in the concept.[94] This approach allows not only the state level and the dynamics within it to be included in considerations, but also the regional level and the importance of the interconnectedness between the different levels.

## Regional and Global Spheres

While power has been discussed earlier in this chapter, it is important to highlight the significance of the regional level for this study. The traditional notion of 'national security' (defined as protection of the state's sovereignty and territorial integrity from external military threats) is still of major importance. However, states – as well as non-state actors – have also raised broader, contemporary concerns that go well beyond external military threats, including human and environmental security. Post-Cold War security has a more local and regional focus, and, therefore, places more emphasis on transnational threats and transboundary cooperation.

Jens-Uwe Wunderlich argues that the regional level gained importance throughout the 1990s because of the worldwide globalisation process and a stronger interconnection between states and regions.[95] This assertion is also supported by Barry Buzan, who argues that the collapse of the Cold War order, a bipolar system and the subsequent decentralisation of the international system have strengthened regionalism.[96] The removal of superpower rivalry has encouraged multipolarity and contributed to a system in which regional agreements can have greater importance. Thus, through the emergence of globalisation, regionalism has gained in significance and is currently the focus at both water resource management theory and policy.

Another fundamental element of new approaches to regionalism is that state actors are not the only actors relevant to the process. Moreover, new regionalism includes interactions between interstate and global institutions, and incorporates the roles of non-state actors such as multinational corporations, emerging civil societies and NGOs.[97] According to Frederik Söderbaum, one of the main characteristics of the new regionalism is a range of formal and informal triangular relations, including a variety of states and non-state actors (private companies and civil societies) in a relatively informal setting that developed after the bipolar world order.[98] Helga Haftendorn takes this further and analyses how issue-specific, regional and global institutions are useful in constraining, resolving or preventing freshwater conflicts. She uses the example of the conflict between Brazil, Paraguay and Argentina on the Parana River over power stations. This incident resulted in a provision by the UN to provide more information and consultation, which was later encompassed in the regional framework of the Southern Cone Common Market (Mercosur).[99] Haftendorn also links this to a level of interaction and applies it to the EU. Achieving consensus over the risk and costs of a settlement is easier in a cooperative environment than between riparians lacking a tradition of solving conflict. Expanding on this, Haftendorn argues that the existence of a facilitator is helpful, as evidenced by the use of the International Court of Justice (ICJ) in the case of the Gabcikovo–Nagymaros project on the Danube River.

Besides the regional level, there are two main global conventions that aim to establish general principles for the use of transboundary water resources. The first is the 1992 Convention on the Protection and Use of Transboundary Watercourses and International Lakes, also called the Helsinki Convention or UNECE (United Nations Economic Commission for Europe) Water Convention.[100] The Helsinki Convention was originally negotiated as a regional instrument; however, the convention was amended in 2003 to allow entry for all UN member states. Amendments effective from February 2013 turned the convention into a global legal framework for transboundary water cooperation, and non-Economic Commission for Europe (ECE) countries could join it from late 2015.[101] However, the UN Water Convention does not include a supranational element or the possibility of sanctions and is limited to the codification of international customary rights. As a result of the structure and expansion of the UNECE membership, the success of the convention as a global framework is yet to be seen.

The second global convention is the 1997 UN Convention on the Law of the Non-Navigational uses of International Watercourses. This convention entered into force in August 2014 with the 35th country ratifying, accepting or approving the convention.[102] However, delays in the ratification process in the last 15 years shows that the management of transboundary water courses is not yet a priority area in the world's political agenda and that riparians prefer to have bilateral or multilateral negotiation between riparian states instead of ratifying a global convention.[103] However, this might change in the next few years if both of the conventions keep up the momentum and gain more national and regional support. Of the three empirical case studies analysed in this book, only few riparians out of 30 signed the UN Water Convention or the UN Convention on non-navigational uses. This supports the view that in 2015 in regards to transboundary water resource management, RBOs and regional frameworks are more important and influential than the current global frameworks.

## Developing a Research Framework

As highlighted earlier in this chapter, in international relations excessive focus is given to military and economic power, and there is an insufficient understanding of what constitutes other types of power. As a result, this research shows that power from a realist perspective, mainly defined in military and economic capabilities, is not enough to explain relations in transnational river basins. Moreover, it is important to include the role of bargaining and issue linkage, as different issues between countries are interdependent and cannot be seen as independent variables. This is also linked to the importance of the regional level, as this is often underemphasised in international relations, especially when focusing on power theories. It is important to include the regional level in order to provide a comprehensive understanding of interlinkages and interdependences between the state level and the international system. As highlighted by Barry Buzan, 'one cannot understand national security of any given state without understanding the international pattern of security interdependence in which it is embedded'.[104] This is especially important in this research, as it provides a framework to understand the position of the states in relation to each other and *vis-á-vis* more powerful states which are crucial to understanding water resource management.

Looking through this regional prism allows a methodical approach to combine the micro level of domestic affairs and states to the macro-level of great power impacts on the international system. This demonstrates the interlinkages of all three levels and how they influence each other, something that is especially important in a transnational issue area. This approach also justifies why certain regions are compared and what impact they have on local and global levels. This is also supported by Mirumachi, arguing that 'analysis must also include scalar considerations of what happens at the domestic, regional and basin-wide level'.[105]

This work also expands on other scholars' work analysing various levels of analysis; including, for instance, Tun Myint's work expanding beyond linking the national and international levels together by linking the local and the international layers of analysis together in his analysis of the Mekong and Rhine rivers.[106] Or Naho Mirumachi's work emphasising the importance of national regulatory and accountability capacities in transboundary river basins between riparian states, but also between different sectors within nation states.[107]

Linking from this, as stated earlier, the book is concerned with both areas of cooperation and conflict, and argues that there are nuances within both in transboundary river basins today. These can range from cooperation provided by notification processes and data exchanges to more sophisticated levels that include institutional designs. Conflict can include, in extreme cases, war and armed conflict; however, it can also manifest in lesser forms over water allocation, hydropower projects or other disagreements. These relations between states can also change over time, meaning that states in a river basin can be cooperative during certain periods and uncooperative in others. My study will focus on the relation and interaction between states and uses the results of three key decisions made within the three selected river basins. Overall, this research will challenge the commonly held assumption that when the upstream country is also the basin's hegemon there is little incentive to cooperate. Based on this, my research asks the following main question: To what extent do power asymmetries prevent or inhibit cooperation between riparian states over water resources?

Developed from this research question, the following four hypotheses emerged from the literature review. The first hypothesis is linked to the statement that two different scenarios exist for hydropolitics in asymmetric power relations. In one, the upstream country is also the basin's hegemon, and there is little incentive to cooperate. In the other, the downstream state is the hegemon and cooperation is possible; however, it depends on and is controlled by the hegemon. This study will focus on the first scenario and will question this assumption. The first hypothesis is that conflict is less likely if multilateral institutions exist, including all riparian states, and if power distribution is equal or with the downstream country. The second hypothesis is that conflict is less likely if the existing multinational institutions have strong legal frameworks, including dispute resolution mechanisms. The third hypothesis is that cooperation is more likely if a sense of regionalism exists. The fourth hypothesis is that cooperation is more likely if there are strong linkages between the states.

To test these hypotheses, and to answer the key research question, this book provides a three-level analysis (namely the regional framework, the RBOs and the micro-level analysis) across the three case studies (see Table 1.1). These three levels are linked to the developed hypotheses, where hypothesis one – that conflict is less likely if multilateral institutions include all riparian states – is linked to the second level of analysis, the RBOs. Hypothesis two focuses on the second and third level of analysis, that is, the institutions and the micro-level analysis. The third hypothesis is linked to the first level of analysis – the regional framework – and the last hypothesis addresses all three levels of analysis and their linkages.

Moreover, the three selected case studies – the Mekong, Danube and La Plata river basins – all support the three-level analysis. The cases were selected as they each have water-related security challenges; a dominant power that is an upstream country; and an existing institution dealing with water resource management, namely the Mekong River Commission (MRC); the International Commission for the Protection of the Danube River (ICPDR); and the Comité Intergubernamental Coordinador de los Países de la Cuenca del Plata (CIC).

Further, each of these rivers are over 2,500 kilometres long, rank in the top 20 rivers worldwide and, together, comprise 30 countries. The study will compare each case and investigate whether the results vary depending on the regional context and the structure of the institution. Concluding from this, this book argues that cooperation in transboundary river basins is possible in asymmetric power relations and that the argument that military and economic power is the only driver behind cooperation and decision-making has limited utility.

The research shows that issue linkage and bargaining play a fundamental role, as does the influence of the regional umbrella – including historical ties, institutional design and political system – in which the basin is located.

## Notes

1 Aaron T. Wolf, "Conflict and Cooperation Along International Waterways," *Water Policy* 1, no. 2 (1998): 251.
2 Task Force on Transboundary Waters, "Transboundary Waters: Sharing Benefits, Sharing Responsibilities," UNESCO, Zaragoza, Spain, 2008, 8.
3 Peter H. Gleick, "Water and Conflict: Fresh Water Resources and International Security," *International Security* 18, no. 1 (1993): 84–5; Frederick W. Frey, "The Political Context of Conflict and Co-Operation over International River Basins," *Water International* 18, no. 1 (1993): 60.
4 United Nations (UN), "Convention on the Protection and Use of Transboundary Watercourses and International Lakes," UN, Helsinki, 1992, Article 1.
5 Lilian Del Castillo, "The La Plata Basin System against the Background of Other Basin Organizations," *International Journal of Water Resources Development* 27, no. 3 (2011): 513.
6 Frey defines 'international rivers' as rivers that form the boundary/boundaries between two or more nations and 'transnational rivers' as rivers that flow across international boundaries creating upstream and downstream riparians. Frey, "The Political Context of Conflict and Co-Operation over International River Basins," 55.
7 Arun P. Elhance, *Hydropolitics in the 3rd World – Cooperation and Conflict in International River Basins* (Washington, DC: United States Institute for Peace, 1999), 3.
8 Jesse H. Hammer and Aaron T. Wolf, "Patterns in International Water Resource Treaties: The Transboundary Freshwater Dispute Database," *Colorado Journal of International Environmental Law and Policy* 9 (1998): 159.
9 Gleick, "Water and Conflict: Fresh Water Resources and International Security."
10 Miriam R. Lowi, "Water and Conflict in the Middle East and South Asia: Are Environmental Issues and Security Issues Linked?," *Journal of Environment and Development* 8, no. 4 (1999): 380.
11 Frey, "The Political Context of Conflict and Co-Operation over International River Basins."
12 Gleick, "Water and Conflict: Fresh Water Resources and International Security," 84–5.
13 This is besides other political tensions which could arise between both countries due to their tense relationship.
14 Gleick, "Water and Conflict: Fresh Water Resources and International Security."
15 Herodotus is a famous Greek historian. Quoted in Ana Elisa Cascao, "Ethiopia – Challenges to Egyptian Hegemony in the Nile Basin," *Water Policy* 10, no. 2 (2008): 19.
16 Marwa Daoundy, "Asymmetric Power: Negotiating Water in the Euphrates and Tigris," *International Negotiation* 14, no. 2 (2009): 370.
17 Wolf, "Conflict and Cooperation Along International Waterways."
18 Ibid.
19 Ibid.
20 Asti K. Biswas, "Cooperation or Conflict in Transboundary Water Management: Case Study of South Asia," *Hydrological Sciences Journal* 56, no. 4 (2011a): 667.
21 Ramiro Berardo and Andrea K. Gerlak, "Conflict and Cooperation Along International Rivers: Crafting a Model of Institutional Effectiveness," *Global Environmental Politics* 12, no. 1 (2012): 109.

22 Susanne Schmeier, Andrea K. Gerlak, and Sabine Schulze, "Who Governs Internationally Shared Watercourses? Clearing the Muddy Waters of International River Basin Organizations." Earth System Governance Working Paper No. 8, 2013, 9.
23 Berardo and Gerlak, "Conflict and Cooperation Along International Rivers: Crafting a Model of Institutional Effectiveness."
24 Thomas Bernauer, "Explaining Success and Failure in International River Management." *Aquatic Science* 64, no. 1 (2002): 1–19.
25 Mark Zeitoun and J. A. Allan, "Applying Hegemony and Power Theory to Transboundary Water Analysis," *Water Policy* 10, no. 2 (2008): 41; Steven Lukes, *Power: A Radical View* (Hong Kong: MacMillan Press, 1974).
26 Naho Mirumachi, *Transboundary Water Politics in the Developing World* (New York: Routledge, 2015).
27 Claudia W. Sadoff and David Grey, "Beyond the River: The Benefits of Cooperation on International Rivers," *Water Policy* 4, no. 5 (2002): 401.
28 Ibid., 401.
29 Ines Dombrowsky, *Conflict, Cooperation and Institutions in International Water Management* (Cheltenham: Edward Elgar, 2007).
30 Frey, "The Political Context of Conflict and Co-Operation over International River Basins," 57.
31 Zeitoun and Allan, "Applying Hegemony and Power Theory to Transboundary Water Analysis," 6.
32 Mark Zeitoun and Naho Mirumachi, "Transboundary Water Interaction I: Reconsidering Conflict and Cooperation," *International Environmental Agreements* 8, December (2008): 297–9.
33 Thomas Hobbes, *Leviathan* (Harmondsworth: Penguin Classics, 1986), 184.
34 Robert Keohane and Joseph Nye, "Interdependence in World Politics," in *The Theoretical Evolution of International Political Economy: A Reader*, eds. George T. Crane and Abla Amawi (New York: Oxford University Press, 1997), 23–37.
35 Gleick, "Water and Conflict: Fresh Water Resources and International Security."
36 Hans Morgenthau, *Politics among Nations: The Struggle for Power and Peace*, 4th ed. (New York: Knopf, 1948), 25.
37 Max Weber, *The Theory of Social and Economic Organization* (Chicago, IL: Free Press, 1947), 152.
38 Robert A. Dahl, *Who Governs? Power and Democracy in an American City* (New Haven, CT: Yale University Press, 1961), 2.
39 Robert A. Dahl, "The Concept of Power," *Behavioral Science* 2, no. 3 (1957): 202–3.
40 Quoted in Chris Brown and Kirsten Ainley, *Understanding International Relations*, 3rd ed. (New York: Palgrave Macmillan, 2005), 98.
41 Jack Snyder, "One World, Rival Theories," *Foreign Policy Journal* November/December, no. 145 (2004): 54.
42 John J. Mearsheimer, *The Tragedy of Great Power Politics* (New York: W. W. Norton & Company, 2001), 361.
43 Ibid., 5.
44 Annette Freyberg-Inan. *What Moves Man* (Albany, NY: State University of New York Press, 2004), 65.
45 Kenneth N. Waltz, *Theory of International Politics* (New York: Random House, 1979); Natalie Kempkey, Margaret Pinard, Victor Pochat, and Ariel Dinar, "Negotiations over Water and Other Natural Resources in the La Plata River Basin: A Model for Other Transboundary Basins?," *International Negotiation* 14, no. 2 (2009): 260.
46 Daoundy, "Asymmetric Power: Negotiating Water in the Euphrates and Tigris."
47 Gleick, "Water and Conflict: Fresh Water Resources and International Security."
48 Robert Gilpin, *War and Change in World Politics* (New York: Cambridge University Press, 1981).

49 John Ikenberry and Michael Mastanduno, "International Relations Theory and the Search for Regional Stability," in *International Relations Theory and the Asia–Pacific*, eds. John Ikenberry and Michael Mastanduno (New York: Columbia University Press, 2003), 2–5.
50 Miriam R. Lowi, *Water and Power: The Politics of a Scarce Resource in the Jordan River Basin* (Cambridge: Cambridge University Press, 1993), 169.
51 Miriam R. Lowi, "Rivers of Conflict, Rivers of Peace," *Journal of International Affairs* 49, no. 1 (1995): 131.
52 Shlomi Dinar, "Power Asymmetry and Negotiations in International River Basins," *International Negotiation* 14, no. 2 (2009a): 355.
53 Lowi, *Water and Power: The Politics of a Scarce Resource in the Jordan River Basin*; Dinar, "Power Asymmetry and Negotiations in International River Basins."
54 Mearsheimer, *The Tragedy of Great Power Politics*, 5.
55 Lowi, *Water and Power: The Politics of a Scarce Resource in the Jordan River Basin*.
56 Zeitoun and Allan, "Applying Hegemony and Power Theory to Transboundary Water Analysis"; Lukes, *Power: A Radical View*.
57 Cascao, "Ethiopia – Challenges to Egyptian Hegemony in the Nile Basin."
58 Ibid.
59 Frey, "The Political Context of Conflict and Co-Operation over International River Basins."
60 Stefano Guzzini, "Structural Power: The Limits of Neorealist Power Analysis," *International Organization* 47, no. 3 (1993), 452–3.
61 Thomas C. Schelling, *The Strategy of Conflict* (Harvard, MA: Harvard University Press, 1980); Ariel Rubinstein, "Perfect Equilibrium in a Bargaining Model," *Econometrica* 50, no. 1 (1982): 98.
62 Robert Powell, "Bargaining Theory and International Conflict," *Annual Review of Political Science* 5 (2002): 15.
63 Cascao, "Ethiopia – Challenges to Egyptian Hegemony in the Nile Basin," 20.
64 Daoundy, "Asymmetric Power: Negotiating Water in the Euphrates and Tigris," 365.
65 Joseph S. Nye, *Soft Power* (New York: Public Affairs, 2004).
66 Dinar, "Power Asymmetry and Negotiations in International River Basins," 332.
67 Oran R. Young, "The Politics of International Regime Formation: Managing Natural Resources and the Environment," *International Organization* 43, no. 3 (1989): 359, 368, 374.
68 Ernst B. Haas, "Why Collaborate? Issue-Linkage and International Regimes," *World Politics* 32, no. 3 (1980): 362.
69 Gleick, "Water and Conflict: Fresh Water Resources and International Security," 82.
70 Marc A. Levy, Peter M. Haas, and Robert O. Keohane, "Institutions for the Earth: Promoting International Environmental Protection," *Environment* 34, no. 4 (1992): 14.
71 Young, "The Politics of International Regime Formation: Managing Natural Resources and the Environment," 365.
72 Haas, "Why Collaborate? Issue-Linkage and International Regimes," 371–2.
73 The PKK is a militant left-wing organisation based in Turkey and Iraqi Kurdistan. The PKK fights for cultural and political rights and more autonomy for the Kurds in Turkey.
74 Daoundy, "Asymmetric Power: Negotiating Water in the Euphrates and Tigris."
75 Helga Haftendorn, "Water and International Conflict," *Third World Quarterly* 21, no. 1 (2000): 53.
76 See Table 1 in Wolf, "Conflict and Cooperation Along International Waterways," 257; Hammer and Wolf, "Patterns in International Water Resource Treaties: The Transboundary Freshwater Dispute Database."

77 The convention was triggered by the Alsatian mines incident in north-eastern France; Tun Myint, "Adaptability of International River Basin Regimes: Linkage Problems in the Rhine," in *Adaptive and Integrated Water Management*, eds. Claudia Pahl-Wostl, Pavel Kabat, and Joern Moeltgen (Berlin: Springer-Verlag, 2008), 131.
78 Daniel Deudney, "The Case against Linking Environmental Degradation and National Security," *Millennium – Journal of International Studies* 19, no. 3 (1990): 471.
79 Stephen D. Krasner, *Structural Conflict: The Third World against Global Liberalism* (Berkeley, CA: University of California Press, 1985); Robert Gilpin, *The Political Economy of International Relations* (New York: Princeton University Press, 1987).
80 Joseph M. Grieco, "Anarchy and the Limits of Cooperation: A Realist Critique of the Newest Liberal Institutionalism," *International Organization* 42, no. 3 (1988): 487.
81 Waltz, *Theory of International Politics*, 162.
82 Morgenthau, *Politics among Nations: The Struggle for Power and Peace*, 512; Waltz, *Theory of International Politics*, 116.
83 Charles Glaser, "Realists as Optimists: Cooperation as Self-Help," *International Security* 19, no. 3 (1995): 55; John J. Mearsheimer, "The False Promise of International Institutions," *International Security* 19, no. 3 (1995): 5.
84 Mearsheimer, "The False Promise of International Institutions," 5.
85 Shlomi Dinar, "Scarcity and Cooperation Along International Rivers," *Global Environmental Politics* 9, no. 1 (2009b): 110; Lowi, *Water and Power: The Politics of a Scarce Resource in the Jordan River Basin*.
86 Kempkey *et al.*, "Negotiations over Water and Other Natural Resources in the La Plata River Basin: A Model for Other Transboundary Basins?."
87 Robert Keohane, *After Hegemony* (Princeton, NJ: Princeton University Press, 1984).
88 Levy *et al.*, "Institutions for the Earth: Promoting International Environmental Protection," 13.
89 Ikenberry and Mastanduno, "International Relations Theory and the Search for Regional Stability."
90 Levy *et al.*, "Institutions for the Earth: Promoting International Environmental Protection."
91 Keohane, *After Hegemony*, 9.
92 Frey, "The Political Context of Conflict and Co-Operation over International River Basins," 57.
93 These three treaties include the La Plata Treaty, the Treaty of the Financial Fund for the Development of the La Plata Basin (FONPLATA) and the Waterway Agreement.
94 Barry Buzan, *People, States and Fear*, 2nd ed. (New York: Harvester Wheatsheaf, 1991), 189–90, 196–7.
95 Jens-Uwe Wunderlich, "Regionalism and Integration Theory. The First Wave: Traditional Approaches," in *Regionalism, Globalisation and International Order*, ed. Jens-Uwe Wunderlich (Hampshire and Burlington: Ashgate, 2007), 2.
96 Barry Buzan, "The Interdependence of Security and Economic Issues in the New World Order," in *Political Economy and the Changing Global Order*, eds. Richard Stubbs and Geoffrey Underhill (New York: St. Martin's Press, 1994), 94–5.
97 Shaun Breslin and Richard Higgott, "Studying Regions: Learning from the Old, Constructing the New," *New Political Economy* 5, no. 3 (2000): 347.
98 Frederik Söderbaum, "Introduction: Theories of New Regionalism," in *Theories of New Regionalism*, eds. Frederik Söderbaum and Timothy M. Shaw (Hampshire and New York: Palgrave Macmillan, 2003), 1–2.
99 Haftendorn, "Water and International Conflict," 53.
100 UN, "Convention on the Protection and Use of Transboundary Watercourses and International Lakes."

101 The UNECE was set up in 1947 and is one of the five regional commissions of the UN. It includes 56 states from Europe, North America and Asia. For all member states see www.unece.org/oes/nutshell/member_States_representatives.html
102 UN, "Convention on the Law of the Non-Navigational Uses of International Watercourses," https://treaties.un.org/Pages/ViewDetails.aspx?src=IND&mtdsg_no=XXVII-12&chapter=27&lang=en (accessed 13.10.2014).
103 Biswas, "Cooperation or Conflict in Transboundary Water Management: Case Study of South Asia."
104 Buzan, *People, States and Fear*, 187.
105 Naho Mirumachi, *Transboundary Water Politics in the Developing World* (New York: Routledge, 2015), 6–7.
106 Tun Myint, *Governing International Rivers – Polycentric Politics in the Mekong and the Rhine* (Cheltenham: Edward Elgar, 2012); Myint, "Adaptability of International River Basin Regimes: Linkage Problems in the Rhine," 133.
107 Naho Mirumachi, "Transboundary Water Security: Reviewing the Importance of National Regulatory and Accountability Capacities in International Transboundary River Basins." Department of Geography, King's College London Working Paper No. 64, 2013, 1–24.

# 3 Troubled Mekong: Upstream–Downstream Challenges and Current Developments

This first case study introduces the Mekong River as an important subregion (politically, economically and ecologically) in the Asia–Pacific that has implications for the wider region. The chapter starts with a brief historical overview before focusing on the stakes of the individual riparians. This section also mentions existing disputes and challenges in other issue areas that can impact on the decisions made regarding water resource management. The second part focuses on current multilateral cooperation in the Mekong region and highlights each institution's specific goals, aims, members, donors, actions and implications for the region. This includes the Association of Southeast Asian Nations (ASEAN) as a wider regional umbrella for development in the region, before the focus is shifted to the Mekong River Commission (MRC) and the Greater Mekong Subregion (GMS) and their influence in the region, including how they have addressed environmental challenges and sustainable development over the last decade. Lastly, the chapter uses case studies within the basin to provide an insight into decision-making on three key decisions made in recent times. These include the case of the Xayaburi Dam in the north-western part of Laos; the Lower Sesan II (LS II) tributary dam in Cambodia; and the blasting of rapids in northern Laos to improve navigation between Yunnan Province and Thailand via Laos.

The key findings of the Mekong River case study are that although the principal purposes of the GMS cooperation are resource development and infrastructure, the GMS plays a pivotal role in keeping political stability and security in the Mekong subregion, with wider repercussions in the Asia–Pacific region. In the last decade in particular, the GMS has developed a focus on environmental issues and can be the driver of future successful cooperation in the Mekong subregion. The MRC is an important body undertaking research and data collection on water resource management. However, it is limited in its mandate, which often weakens the riparian body and limits its success. The Procedures for Notification, Prior Consultation and Agreement (PNPCA) process and its first-time use further weakens the MRC's ability to influence the riparians with their findings and recommendations regarding dam developments on the mainstream. While ASEAN is an umbrella for the subregion, environmental matters and guidelines do not influence the subregion strongly as the procedures and guidelines are often implemented slowly and not thoroughly monitored, in contrast to the Danube River Basin with the European Union (EU) Water Framework Directive. This is especially showcased through the micro cases.

## Importance of the Mekong Subregion

As the largest river in Southeast Asia and the eighth largest[1] in the world, the Mekong provides enormous economic and ecological resources as well as a political significance that extends to the six riparian countries.[2] The Mekong is an important subregion in the Asia–Pacific as more than 70 million people live directly on its river banks (see Map 3.1). Additionally, the river hosts a unique and significant ecological system with some of the world's highest diversity of fish and snails, including the endangered Irrawaddy dolphin and the giant Mekong catfish. Moreover, today, the capitals of Cambodia and Laos, Phnom Penh and Vientiane are strategically centred on the river.

The Mekong region has been the centre of development and human civilisation for hundreds of years. The Khmer civilisation, which was vitally linked to the Mekong River, centred around Angkor,[3] with a peak of around 1,000 square kilometres in the 1400s.[4] At the end of the nineteenth century, the Mekong River attracted early interest from European powers, especially France, whose explorers discovered the Mekong between 1866 and 1868 in search of new trading routes and a navigational route along the Mekong from Vietnam to Yunnan in China. The Mekong River has been at the centre of human conflict and wars for several hundred years.

In future decades, the Mekong will have even more strategic importance as it is the centre for food, accommodation and employment, and plays a crucial part in the future development of the six riparian countries China, Myanmar,[5] Thailand, Cambodia, Vietnam and Laos. Some of these countries are the poorest in East Asia, with Laos and Cambodia having a per capita Gross Domestic Product (GDP) estimated in 2014 at 5,000 United States Dollars (USD) and 3,300 USD, respectively.[6] River development is currently at a crossroad: it can lead towards peace, cooperation and harmony, or it can potentially lead in the direction of conflict, dispute, insecurity and problems. Floods, droughts, famines, environmental disaster, soil erosion and deforestation are influencing the livelihoods of those living directly on the river. Transnational challenges and environmental degradation affect East Asian countries in different aspects and locations, and are impairing the Mekong subregion in particular.

## What Is at Stake for the Riparian States?

The Mekong River's source is in the Tibetan Plateau – the Rupsa-la-Pass of eastern Tibet – at a height of 5,100 metres; it then drops significantly through the Yunnan Province in southern China. Around 50 per cent of its length is through China where it drops approximately 4,500 metres, which makes it a

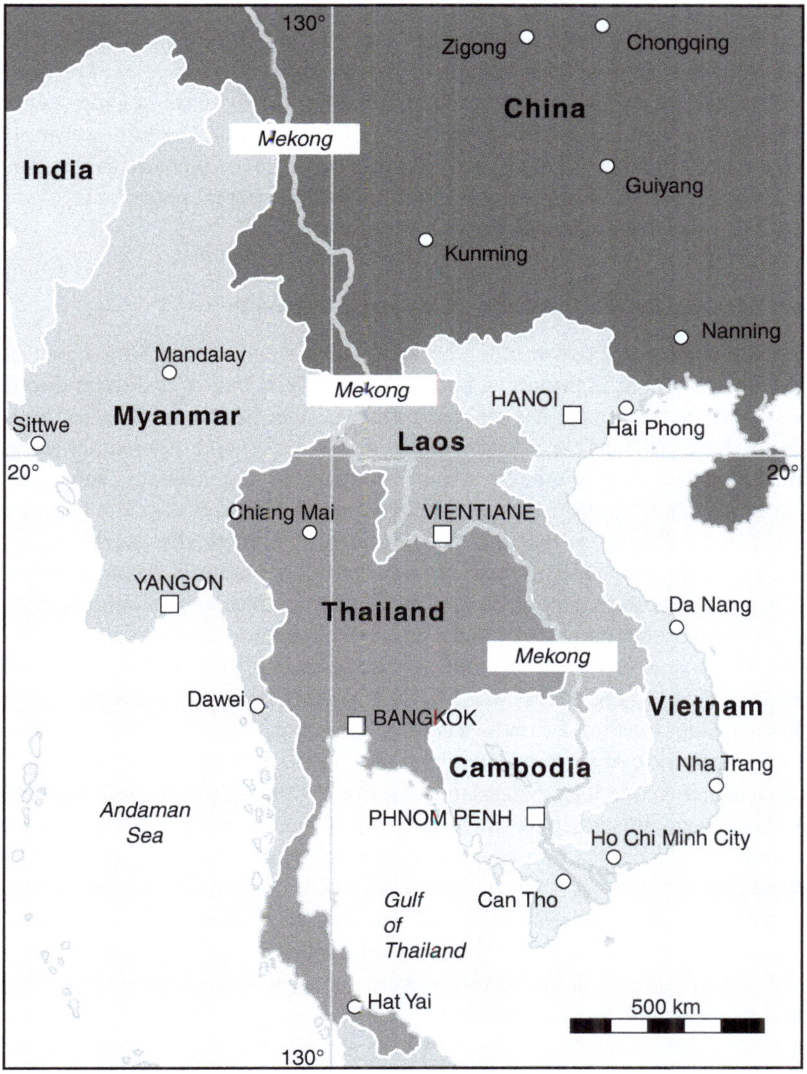


*Map 3.1* The Mekong Subregion.

Source: Adjusted from Haefner, 2013a.

significant potential source of hydropower development in China, the most upstream Mekong country.[7] Leaving China, the Mekong flows through five more countries, covering nearly 4,900 kilometres before ending in the South China Sea at the southern part of Vietnam.[8] The Mekong River is divided into the Upper Basin – consisting of China and Myanmar, which comprises around

24 per cent of the total river water catchment – and flows into the Lower Basin including Laos, Thailand, Cambodia and Vietnam.[9]

The Mekong region is facing a steady increase in hydropower projects as a result of rapid economic development of the riparians in recent years. The need for cheap and renewable energy is rising, fostering the increase in hydropower development where logistically possible. As of mid-2013, 35 hydropower projects have been finalised in the Lower Mekong Basin, all of them on tributaries. Besides hydropower development, each riparian has different plans for its share of the river, which can sometimes be contradictory.

### *Upper Mekong Basin Countries – China and Myanmar*

China is the dominant power in the Mekong Region and around 50 per cent of the Mekong's length is through China. China's role is one of historical dominance in the region and it has exerted influence over Southeast Asia for hundreds of years. China is economically, strategically and demographically dominant over other Southeast Asian countries in the Mekong subregion through links in migration, business networks and capital flows. The Chinese White Paper 2002 states that the Mekong area is strategically important for China's security, especially in regard to non-traditional security threats but also as a boost for western provinces and as a link to continental Southeast Asia through major connecting roads and improved navigation on the river.

The Mekong River is an important strategic and political instrument since it allows China to gain access to ASEAN and other regional developments in Southeast Asia. Further, China's overall economic development also depends on the contribution of the western inland provinces, such as Yunnan, to reduce poverty and increase living standards. Through a closer connection with the cross-border countries, these provinces could increase their trade outside China and could benefit enormously from navigational developments on the Mekong River.

China's behaviour in relation to the Mekong is often seen from different perspectives. Some argue that China will not cooperate to help other countries on the Lower Mekong benefit from natural resource management. However, others, such as Milton Osborne, argue that the events and actions that happened, for instance, on the Nu River in 2006 showed that a growing number of Non-Governmental Organisations (NGOs) can influence the government. This example resulted in the downscaling of 13 planned dams on the Nu River to only four.[10] China's Premier Wen Jinbao in 2004 halted many large dam projects due to environmental impact studies.[11] However, it seems this policy changed again in 2013. China's State Council shows an increasing interest in transboundary rivers as the internal rivers are mostly saturated with dams. It is also important to remember the different views that exist regarding dam developments. Yunnan is preoccupied with damming and is using as many resources as possible to take advantage and expand routes to Thailand and the West.[12] Also, China has not joined the MRC and is unlikely to do so in the future.

The same applies to Myanmar, which is the second country in the Upper Mekong Basin and currently not a signatory to the MRC.

Myanmar is the second largest country in the Southeast Asian region and is rich in natural resources, such as land, forestry, minerals (including gas and oil), freshwater and marine resources. However, due to past decades of being a 'locked up' country before the political and economic reforms in 2011, the country plays so far a minor role in regards to developments on the Mekong River. This is enhanced by the relatively small section of Myanmar on the Mekong alongside the border with Laos on its eastern side. This area is currently less important for the country than other major centres. Myanmar's position and interest will develop over time and in which direction the country develops remains to be seen. Currently, Myanmar builds dams in its northern regions that are funded and developed by Chinese companies and investors; for example, the controversial 3.6 billion USD Myitsone Dam was proposed in Kachin state on the Irrawaddy River, however, is currently halted (as of November 2015).[13]

### *Lower Mekong – Laos, Thailand, Cambodia and Vietnam*

Laos, as one of the poorest countries in the region with a per capita GDP estimated in 2014 of 5,000 USD, focuses solely on increasing its development and the improvements of livelihoods.[14] Revenues from mining and hydropower in Laos have 'skyrocketed' in the last five to ten years.[15] Laos's future economic growth is closely linked to hydropower development due to its beneficial geographic positioning on the Mekong. This advantageous position means sufficient energy not only for its own demand but also as an important source of export revenue as its neighbours, including Thailand, China and Vietnam, face increasing energy shortages.

Dam building in Laos started in the 1970s, and includes the Nam Ngum Dam and the Theun Hinboun Dam (completed in 1998) and is today one of the most pressing topics in the country. The Lao government views hydropower development and especially the Xayaburi dam (the first mainstream dam on the Lower Mekong) as revenue, a source of foreign investment and potential economic growth. Laos wants to become the 'battery of the region', and it is likely to benefit through large construction projects. Besides dams, big construction projects are still seen as a sign of power and a symbol of a growing Laos, which was previously often neglected in the region. Large infrastructure developments are also used to further the power of the current leadership and for the ruling party to become more legitimate. Although there are several critics of the current developments regarding hydropower in Laos (see Xayaburi case study below), the Lao government remains undeterred, stating that there are no other satisfactory options to replace the income of hydropower development. In 2011, Laos exported nearly 678 million kilowatt hours (kWh) electricity (nearly one-third of its total generation), which counted for around one-third of its total exports, showing the importance of hydropower in gaining revenue for the country.[16]

Thailand's interest in the Mekong is positioned in water diversion and irrigation projects to support its arid north-eastern region. The Thai name for the Mekong, 'Mae Nam Khong' or 'Mother of the Waters', highlights the river's importance for its population as a source of food and water for irrigation. However, as the Mekong River flows along the border line and not through the middle of the country or the capital, some people argue that Bangkok does not pay as much attention to the Mekong as other countries in the region do. This is different to the Thai provinces that are built along the Mekong where the river plays a pivotal role in everyday life, especially in the eastern rural areas of Thailand.

As Thailand has already developed most of its domestic hydropower potential, it seeks to import from neighbouring countries. Also, previous strong opposition to projects, including the Pak Mun Dam,[17] further supports the increasing involvement in hydropower development abroad where civil society is weaker. Interestingly, some energy experts such as Witoon Permpongsacharoen, the Director of the Mekong Energy and Ecology Network, argue that electricity demands are overestimated and that existing energy options in Thailand would be sufficient.[18] In addition to hydropower development, Thailand's interest in water irrigation affects Mekong tributaries through the Kok-Ing-Nan Water Diversion Project which balances the demands for water drawn out of the Choa Phraya River. Thailand's aim is to withdraw 2,200 million cubic metres from the three dammed tributaries; this has already provoked local protests, whilst concerns have also been expressed by Cambodia.[19]

Cambodia and Vietnam are the most downstream countries in the Mekong Basin and will face the biggest impacts in the future. Cambodia is concerned about the Tonle Sap Lake, the biggest lake in Southeast Asia. The lake depends on the Mekong River's enormous water flows in the wet season in order to reverse the direction in the dry season, carrying high amounts of fish and other animals that spawn in the lake during the wet season. This counts for at least up to 60 per cent of the annual protein intake of Cambodia's population.[20] Cambodia further depends on nearly 75 per cent of water originating from outside its borders, which makes water security an important task in national politics and could influence its bargaining power.[21]

On the one hand, Cambodia wants to keep the seasonal flow, including the flood pulse system of the Mekong, as opposed to a regulated flow through dams in order to protect the phenomenon of the exceptional Tonle Sap Lake with its unique ecosystem. On the other hand, Cambodia also plans to build dams within its territory, which could have a devastating impact on the region and on the productivity of the Mekong, as will be explained in the LS II case study. Hydropower is expected to account for 77 per cent of Cambodia's total electricity generation by 2030 in contrast to less than 4 per cent in 2007.[22]

Vietnam is also focusing on reducing damming on the Upper Mekong, since around 50 per cent of its annual rice crop is produced in the Mekong delta, which shows increasing salinisation. Because of the Mekong's river existence and characteristics, Vietnamese farmers can grow three rice crops per year in the Mekong Basin, resulting in Vietnam being the world's second largest rice exporter in 2012.[23] The Vietnamese goal of sufficient dry season flow is one of the main issues regarding damming on the Upper Mekong, because this influences the cultivation of rice and aquatic production and could have an impact on rice prices on a national, regional and global level. However, Vietnam also needs to satisfy its growing energy demand and relies on energy imports from neighbouring countries such as Laos. Additionally, whenever possible (due to the geographical settings) Vietnam builds hydropower dams, mainly on the Dong Nai River and Sesan River, to support the growing national energy demand.[24]

### *Disputes Between Countries/Interconnections*

Aside from the different interests and plans for the Mekong, further complications could prevent regional cooperation. Suspicions still exist between the Mekong countries because of the ideological divide and varying support during the regional wars. This has resulted in small-scale conflicts regarding temple sites and border demarcation between Thailand and Laos, and also Thailand and Cambodia. Further minor conflicts and disputes exist between Laos and Thailand over a few islands in the Mekong River. Thailand and Cambodia still deploy troops around the Preah Viear ancient temple in the border region with increasing problems and resulting refugees in 2008 and 2009. Additionally, in early 2011, soldiers were killed in this area and the inhabitants of the border area were threatened. Furthermore, Vietnamese dams on the Sesan River (which rises in Vietnam and flows to Cambodia) cause negative effects such as low water flow, poor fish catches and property loss in Cambodia, leading to disputes between these two countries.

One of the most significant influences and developments on the Mekong River is the role of China. In particular, China has a major political and economic influence in the Mekong region with important trading and cultural links through Chinese ancestry in the countries of the Lower Mekong. While China shows an increasing interest in Cambodia, its ties with Vietnam seem to be decreasing. Between 2006 and 2011, Chinese investors financed around 46 per cent of all hydroelectricity implemented in Cambodia, Laos and Myanmar, which demonstrates their increasing influence.[25] Competition also exists between China, Thailand and Vietnam for natural resources in Laos and Myanmar.

Other related issues include contract farming, rubber plantations, migration and language in the region, all of which provide additional links between the riparians. Logging and illicit wood trading plays a significant role in the relations between Laos and Vietnam, especially in the border area between Vietnam and southern Laos that is trying to adjust to the increasing demand of raw materials in China and Thailand, but also especially in Vietnam due to a booming wood processing industry.[26] While this illegal trade violates Laos' laws, it is very lucrative for corrupt government officials, businessmen, workers and Vietnamese-based logging companies, including the Vietnamese military and furniture factories.[27] Often, these logging activities are linked to investment in rubber plantations, mining and hydropower projects, as hydropower and mining often foster deforestation, given that large areas of forest need to be cleared before large projects can proceed. This highlights the linkage of different issues. A broader influence further away, the South China Sea dispute, could be another area influencing developments on the Mekong, since China and Vietnam have different viewpoints regarding territorial claims mainly around the Paracel and Spratly Islands, which might be linked to strategic decisions made within the Mekong region.

## Multilateral and Bilateral Cooperation in the Mekong Subregion

Having outlined the historical developments in the Mekong subregion, the following section will focus on multilateral regional institutions. ASEAN is composed of ten countries, and it is the driver for regional development in Southeast Asia and the wider Asia–Pacific region. ASEAN includes all countries in the Mekong subregion, with the exception of China (which is, however, part of the ASEAN Plus Three Cooperation (APT)). It is important to understand the role of ASEAN and its procedures in order to understand subregional developments surrounding cooperation.

Besides ASEAN, several other cooperation bodies exist in the Mekong subregion. These feature different development backgrounds, member countries and focal points covering transportation, food productivity and energy development. For instance, the MRC was re-established in its current form in 1995 and includes the Lower Mekong countries, Thailand, Vietnam, Laos and Cambodia, and is largely financially supported by European countries. The GMS is an important institution that was established by the Asian Development Bank (ADB) in 1992; it focuses mainly on economic development and infrastructure. It includes the Lower Mekong countries as well as the Upper Mekong countries, Myanmar and China.

Besides these forms of multilateral cooperation, several other collaborations exist, such as the ASEAN–Mekong Basin Development Cooperation (AMBDC), the Emerald Triangle Quadripartite Economic Cooperation (QEC), the Forum for the Comprehensive Development of Indochina, the Ayeyawady-Chao Phraya-Mekong Economic Cooperation Strategy and Japan's Forum of Comprehensive Development of Indochina (FCDI).

## The Role of the Association of Southeast Asian Nations (ASEAN)

ASEAN is the main body of regional cooperation in Southeast Asia and also the driver for regionalism in the wider Asia–Pacific region. The institution was formed on 8 August 1967 by Indonesia, Malaysia, the Philippines, Singapore and Thailand during a period of regional tension caused by the Vietnam War and the Cultural Revolution in China. Essentially, ASEAN was an effort to band against external threats, including Communism, and was based on a reactive and negative foundation, unlike the European Coal and Steel Community (ECSC), which was based on a shared regional economic and political vision.

ASEAN was institutionalised in 1976 when the Treaty of Amity and Cooperation was signed and has expanded its membership since then to the now-ten ASEAN members.[28] ASEAN today includes two of the world's richest states, Brunei and Singapore, as well as several of its poorest. In addition, politically, ASEAN membership ranges from functioning democracies to authoritarian regimes.

### *ASEAN Characteristics and Its Limitations*

ASEAN is driven by certain characteristics focusing on an informal, widely consultative, incremental and consensus-based approach towards cooperation and negotiations. The primary principles in this treaty are the following:

> mutual respect for the independence, sovereignty, equality, territorial integrity, and national identity of all nations; the right of every State to conduct its affairs free from external interference, subversion or coercion; non-interference in the internal affairs of one another; settlement of differences or disputes in a peaceful manner; renunciation of the threat or use of force; and effective cooperation among themselves.[29]

In general, the ASEAN way is a 'loosely used concept whose meaning remains vague and contested' which is generally based on the traditional regional practices of consensus-finding and consultation.[30] These characteristics are unique not only to ASEAN but also to its subregions and multilateral cooperation bodies, such as the Asian Regional Forum (ARF) or APT.

Although ASEAN is without permanent institutional structures, it holds annual meetings with foreign ministers, a formal summit of leaders every two years as well as an informal summit. To clarify, although ASEAN does not lack regularised ministerial and bureaucratic sessions, it has resisted the idea of a centralised, formalised and permanent bureaucracy with decision-making authority. The ASEAN way is not merely about the structure and outcome of multilateral interaction, but about the process through which its way is carried out. More specifically, ASEAN multilateralism is process-orientated rather than product-orientated.[31] For instance, the key for cooperative security is inclusiveness, meaning that the dialogue must involve as many relevant principal actors as possible. Furthermore, the ASEAN way combines universal principles of multilateralism with some features of socialisation as well as consensus building and the avoidance of grand institutions.

One of the main principles of the ASEAN way, the principle of non-intervention, was discussed in the aftermath of the global financial crisis and showed a minor shift towards an agreement that some ASEAN influence in national policies is plausible.[32] The proposal of 'flexible engagement' suggested by Thailand and the Philippines in the late 1990s promoted a method that would allow greater ASEAN involvement in domestic issues, which have regional and international implications.[33] However, ASEAN rejected the approach for a 'flexible engagement' and instead accepted the 'milder version' of 'enhanced interaction', which only allows limited involvement. This example indicates that occasionally there are discussions about modifying the ASEAN way. However, despite these discussions, there have been no significant changes in the established principle. According to the ASEAN Secretariat in 2006, 'ASEAN's problem is not one of lack of vision, ideas or action plans. The problem is one of ensuring compliance and effective implementation'.[34] This lack of implementation is also visible in the Mekong subregion's organisations where projects and programs exist, but where the implementation and practical support are missing.

A similar problem could occur with the 2007 established ASEAN Charter. This Charter is supposed to enable ASEAN to integrate the regional economy more effectively and expeditiously, serve as a force for regional peace and stability and cooperate in dealing with common regional problems.[35] For the first time, ASEAN included norms not only for dealing with relations between member states but also impacting on the internal behaviour of states. According to Severino, this shift included norms such as democracy, human rights and fundamental freedoms, constitutional government, rule of law, social justice and good governance, with the stated intention of establishing a regional human rights body.

However, it is important to be mindful that the implementation of the Charter depends on the member states and their initiatives. ASEAN's former Secretary-General, Ong Keng Yong, argues that 'face' is important in ASEAN and that the monitoring and reporting process in the Charter affects this face.[36] Nevertheless, there is no method to measure or evaluate either this behaviour in the Charter or whether a change towards an 'enhanced interaction' in ASEAN is visible. The same lack of a measuring tool also applies to the cooperative bodies growing out of ASEAN, such as the Asia–Pacific Economic Commission (APEC), the ARF or Mekong subregion specific bodies, which are closely linked to ASEAN.

The two major problems ASEAN faces are issues relating to regional identity and the principle of non-interference in other domestic member countries, which often slows down processes or impedes policy implementation. The institutional mechanisms as well as the regional philosophies to deal with internal and external affairs have proven ineffective, and real problem-solving requires more concrete action. The new ASEAN Charter includes norms regarding relations between the member states; however, its implementation is not guaranteed. It is still unclear how far the Charter will improve the conditions of internal state behaviour because no sanctions are evident and results depend on the acting member state. Aside from these challenges, the Southeast Asia region potentially has enormous economic strength and its diverse physical and human resources indicate a high growth potential, as long as the countries of the region can work together to benefit from one another's balancing strengths.[37]

ASEAN has many economic strengths, indicates a still-high growth potential, and engages with several major players in the international system without privileging certain countries. As a result, besides the ongoing challenges of terrorism, political unrest and problems in certain member states, ASEAN members have achieved a position in which they are still the centre of developments in the Asia–Pacific region.[38] ASEAN is the key regional player in organisations such as the ARF and the East Asia Summit (EAS). In addition, ASEAN is seen as a neutral institution in which Japan, China, the United States (US) or India can discuss disputes and problems. The institution retains and grants regional legitimacy to US strategic involvement in the region, and engages and incorporates China as a responsible regional power. For instance, ASEAN follows a bilateral and multilateral engagement with China through APT, or bilateral agreements between an ASEAN country and China. Furthermore, ASEAN has links with South Korea and China through free trade agreements.

## *ASEAN – Environmental Legislation*

In regards to environmental challenges, ASEAN has worked to promote environmental cooperation among its member states, which started with the ASEAN Expert Group on the Environment (AEGE) in 1978. Environmental cooperation can be divided into three phases, with the first phase from 1977 to 1987 characterised by a non-formal web of soft law declarations, resolutions and action plans avoiding non-binding agreements.[39] The second phase (1988–1992) included the ASEAN Strategic Plan on the Environment and placed greater emphasis on environmental protection, commonalities and collective responses. In the third phase, AEGE was expanded in 1989 and became the ASEAN Senior Officials on the Environment (ASOEN) with its subsidiary bodies including the Environmental Ministers' meeting and the ASEAN Secretariat.[40]

Early key focus areas included the management of shared natural resources regarding biodiversity and pollution control, mainly referring to haze pollution through forest fires to acquire agricultural land. The Hanoi Plan of Action from 1999 to 2004 covered 15 areas relating to the environment under Paragraph IV of the Action Plan.[41] Currently, ASEAN environmental cooperation includes water management through the ASEAN Cooperation on the Environment. This focuses on ten priority areas of regional importance as reflected in the Blueprint for the ASEAN Socio-Cultural Community (ASCC) 2009–2015,[42] including global environmental issues, promoting the sustainable management of natural resources, biodiversity and freshwater resources.[43] The environmental challenges faced by ASEAN are linked to the ASEAN culture and principles of non-interference and non-binding agreements. Other problems include the general lack of concrete instruments that are able to translate the regional commitments from the ASEAN level into national policy and its implementation.[44]

Together, the countries of Southeast Asia built a collective of small powers who act as one to achieve greater importance and 'ASEAN is an aspiring major power in East Asia'.[45] Similarly, ASEAN countries have 'social trust' and 'social capital' through several annual meetings between politicians, officials and academics, which bind the region together.[46] This connection and role of ASEAN has positive implications for Cambodia, Laos, Vietnam and Myanmar (CLVM) as they are part of ASEAN and, therefore, can attract more bilateral ties with the major players including investment, trade, cooperation and other developmental aid. Overall, it is important to understand the role and actions of ASEAN, which can be seen as an umbrella organisation, in order to understand the Mekong subregion and its specific institutions and developments. Further, whereas subregions benefit from the wider umbrella of ASEAN, growth triangles such as the Mekong subregion often also generate benefits for the wider region through spill-over effects and cooperation. As a result, it is important to look at the second level of analysis – the institutions directly dealing with water resource management.

## Mekong River Commission (MRC)

There is a long history of management and cooperation on the Mekong River. After World War II, the Economic Commission for Asia and the Far East (ECAFE) was established, with a focus on developing the Asian region. A study of the Lower Mekong River was conducted as one of the first initiatives, resulting in a report in 1952 that concluded the necessity of further intense studies of the transnational river. Advanced studies followed and a Committee for Coordination of Investigations of the Lower Mekong Basin, the Mekong Committee (MC), was established in 1957 lasting until 1995 when it was replaced by the MRC.

### *Mekong Committee (MC) and Interim Mekong Committee (IMC) 1957–1995*

The MC was created in 1957 through the US support of the southern Vietnamese state and the use of the Mekong as a bastion against the Communists. The United Nations (UN) National Security Council stated in 1956 that economic development linked to the river would 'deny the general area of the Mekong River Basin to Communist influence or domination'.[47] The so-called Vietnam War involved Cambodia, Laos and Vietnam and led to the defeat of the US, which supported its allies in South Vietnam, Cambodia and Laos.[48] This resulted in the growing power of Communist governments in all the three countries, with the most dramatic developments in Cambodia under the Pol Pot regime, which had a devastating impact on the country's development (it also impacted on the country's participation in the MC during this period).

The MC was a developmental organisation with a focus on investigating the potential for hydropower, irrigation and flood control. The Committee strongly supported the building of large dams on the Mekong mainstream and also smaller dams on its tributaries, including dams at Sambor in Cambodia or Savannakhet and Pa Mong in Laos. The 1950s and 1960s were characterised by a strong unquestioned support of large dams in the context of other projects, including the Aswan High Dam in Egypt, where countries competed for influence through financial support. This was highlighted by the strong involvement of the US under President Johnson, which was also linked to the political interest during this time.[49]

In the 1970s, the MC was development-orientated and planned several dams along the Lower Mekong River as highlighted in the '1970 Indicative Basin Plan'. This plan included 180 possible projects for the basin development, including seven mainstream dams. A few of the short-term projects were implemented, including tributary dams; however, none of the major projects were realised, except for the Nam Ngum Dam. This dam was built in 1971 in Laos during an increasingly difficult political situation, in which the Mekong was the centre of major strategic importance and bitter fighting. Following this, the Mekong Delta was under Communist rule from 1972 until the end of the Vietnam War in 1975. The Nam Ngum was the 'only truly intergovernmental project achieved' by the MC.[50]

Geopolitical events led to the demise of the MC in 1975, with wars impacting on Laos, Vietnam and Cambodia that resulted in no representatives being appointed to the Committee in 1975. However, a year later, the IMC was set up, comprising Laos, Vietnam and Thailand. Under the Pot Pol regime, Cambodia was unable to participate in such an organisation, so, in April 1977, the other three riparians agreed to the Declaration Concerning the IMC, which resulted in the establishment of the IMC in January 1978. The weakened interim organisation shifted largely to data collection and was only able to study large-scale projects and implemented a few small-scale projects, almost exclusively in Thailand. Further studies and development on the Mekong followed with a new indicative plan and projects, including the Pak Mun Dam in Thailand and the Yali Falls Dam[51] in Vietnam. Cambodia requested readmission to the Committee in 1991, and after further bargaining by Thailand (which wanted to negotiate a new framework), the UN Development Program (UNDP) organised several meetings, resulting in the establishment of the MRC several years later.

### *Mekong River Commission (MRC) since 1995*

In 1995 the IMC was succeeded by the MRC, which included the three previous countries and the newly re-joined Cambodia. All four riparians signed the Agreement on the Cooperation for Sustainable Development on the Mekong River Basin and established the MRC. The central focus of the new commission was to cooperate on the sustainable development, utilisation, conservation and management of the Mekong River Basin water and related resources, including irrigation, hydropower, navigation, flood control, fisheries, timber floating, recreation and tourism. Other goals included to 'optimize the multiple-use and mutual benefits of all riparians and to minimize the harmful effects that might result from natural occurrences and man-made activities'.[52]

This agreement for sustainable development on the Mekong River Basin established the MRC as it is today, consisting of three permanent bodies: the Council, the Joint Committee (JC) and the Secretariat (see Figure 3.1).[53] At the annual Ministerial Council Meeting, one member of each country meets at the ministerial or cabinet level. The JC consists of senior bureaucrats, who provide the organisation board with technical and logistical advice. The Secretariat, led by the Chief Executive Officer (CEO), also plays a crucial role. The MRC

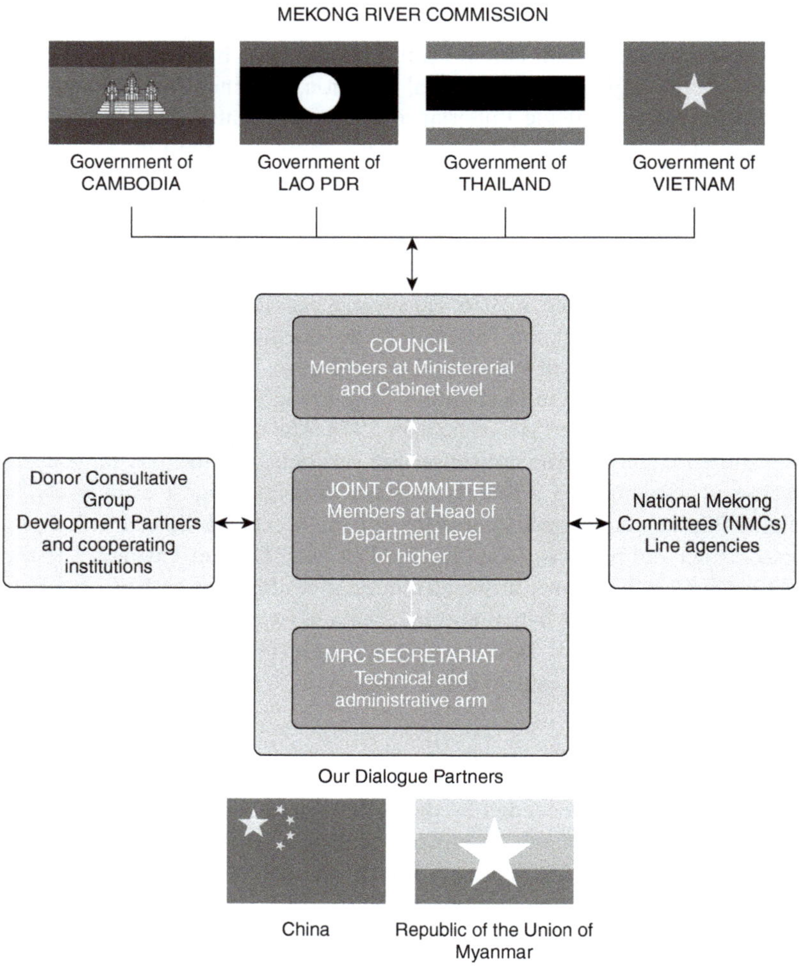


*Figure 3.1* MRC organisational structure.

Source: Adjusted from MRC, 2013.

Secretariat is permanently co-hosted in two locations – the Office of the Secretariat in Vientiane (OSV) and the Office of the Secretariat in Phnom Penh (OSP).

Positive early infrastructure developments in the Mekong subregion include the first bridge across the Lower Mekong in 1994 between Thailand and Laos (the so-called 'Friendship Bridge'), which was funded by the Australian government as development aid to Laos. Other projects included the reopening of the rail link between China and Vietnam in 1996, followed by the bridge between Myanmar and Thailand near Maesot in 1997.[54] As already realised by the early French explorers, the navigation route on the Mekong is still limited, and ocean-going vessels can travel as far as Phnom Penh. Navigation on other river sections by vessels of any size is limited to the rainy season when, for around three months, the water is at its peak.

Although the membership group of the MRC was the same as the MC and the structure of the organisation similar, the MRC's focus in 1995 was different to that of the MC. The MC was founded as an economic cooperation forum focusing on development, whereas the 1995 agreement combines economic development with environmental protection, concentrating on a basin-wide management approach. Since 1995 the MRC has developed a special focus on issues that affect more than one country and is developing 'rules' for water sharing, monitoring the quality of water resources and supporting a Basin Development Plan.[55] For instance, the MRC's Annual Report in 2001 focused on implementing three core programs: the Basin Development Plan, the Water Utilisation Program and the Environment Program. The Annual Report emphasised the importance of 'bottom-up' solutions and the 'voice of the people directly affected', which was a major shift within the MRC. Similarly, the 2001 MRC Hydropower Development Strategy explicitly disavowed the 'promotion of specific projects' in favour of 'basin-wide issues'. In part, these shifts mark a retreat from past project failures, and indicate a recognition that the MRC faces multiple, and often more lucrative, competitors in regards to project management. Some of the more recent initiatives by the MRC include the Initiative on Sustainable Hydropower and the Rapid Basin-wide Hydropower Sustainability Assessment Tool.

### *Members and Funding*

Since the early 1990s, the MRC secretary has stressed the importance of including the upstream countries in coordinating the Mekong River. In 1996, China and Myanmar became dialogue partners of the MRC and slowly expanded their (non-binding) participation in various forms, including data exchange and scientific research. Since early 2002, China has sent technical daily water level and rainfall data to the MRC during the rainy season in order to forecast possible flooding downstream. However, reliable and continuous data exchange and water release schedules from Chinese reservoirs during the dry period still do not exist; although in the 2010 dry season China began to provide data from the Jinghong Power Station on the mainstream Mekong and the Man'an tributary for downstream countries, which further contributes to more transparency.[56]

This changing behaviour can be seen as an indicator of China's changing attitude towards Southeast Asia. However, as only dialogue partners, China and Myanmar do not need to follow the decisions or agreements made by the MRC. This also counts in regards to the PNPCA or notification processes regarding new infrastructure projects affecting the Mekong water flow. At present, it is unlikely that either country will join the MRC in the near future. Overall, while the MRC is an international organisation between the four downstream countries, until recently it was mainly foreign donors who play a crucial role in its funding, which is seen as problematic because of the very low local ownership and acceptance. In 2014, around 80 per cent of the MRC funding came from external donors, including the Netherlands, Australia, Sweden, Germany, Finland and the ADB; whereas around 20 per cent came from the riparian countries.[57] However, the new Strategic Plan 2016–2020 includes an annual 10 per cent increase of the financial contributions from the member countries to reduce the reliance on financing from donor countries with the goal of achieving self-financing by 2030.[58]

### *The Role of the MRC Chief Executive Officer (CEO)*

Besides the discussed role of the membership and the funding, the role of the MRC CEO (leading the MRC Secretariat) also plays a crucial part in the direction and focus of the organisation. Yasunabu Matoba, a Japanese professional irrigation engineer, was the first CEO of the MRC. Focusing strongly on large-scale programs often funded through Japanese aid, the newly established MRC followed in the footsteps of the MC. Joern Kristensen, however, during his term as CEO from 2000 to 2003 identified the MRC mission as ensuring sustainable water and resource management. Under his leadership, publications such as the 'State of the Basin Report 2003', the 'Social Atlas of the Lower Mekong Basin' and the 'River Awareness Kit' were published.[59]

By contrast, Olivier Cogels, CEO from 2004 to 2007, promoted a new direction for the Mekong region – supporting economic development and aiming to reduce the environmental focus of the organisation. Cogels argued that the negative impacts of the dams had been exaggerated and that dams help with water control in the dry season since they reduce flooding. For example, in a letter to the *Bangkok Post* in 2007, Cogels argued that Chinese dams would increase the river's dry season volume as their purpose was electricity generation and not irrigation.[60] Undoubtedly, dams can increase dry season flow; however, there is uncertainty about Chinese reservoir policies and dam purpose during the dry season. Since China is not included in the downstream cooperation of the MRC, it does not have any guidelines on reporting water levels or water release during the dry season. Echoing Cogels, public statements from MRC leaders have earned the MRC a reputation of being responsible for accepting Chinese dam building. However, the Nam Ngum Dam in Laos is the only hydropower project supported by the Commission and it shows the changing policies of the organisation and its leaders.

Jeremy Bird, who followed Cogels as CEO from 2008 until 2011, focused on the pressure placed on the resources in the Mekong River, mainly concentrating on water resource management. During his term, special attention was directed to social and environmental impact assessments that relate to big hydropower plans.[61] Under his leadership, the 2010 Strategic Environmental Assessments were made, highlighting the impact of mainstream damming on the communities, fisheries and hydropower on the river.[62] Bird further emphasised the importance of a cooperative, inclusive approach in order to achieve water resource management.

Hans Guttmann, a Swedish national, was nominated at the end of 2011 and was in office until early 2015. Guttmann knows the region and the organisation well, as he was the head of the Environment Program between 2001 and 2007.[63] His appointment was followed by Sweden's reaffirmed investment in the Mekong region with a focus on the Lower Mekong Basin. From 2012 to mid-2015 Sweden expanded its support for the Basin Development Plan Program, the Environment Program, the Fisheries Program and institutional operations, such as communications, stakeholder engagement and the decentralisation of the MRC's core functions.[64] Since March 2016, Pham Tuan Phan of Vietnam is the first riparian CEO since the foundation of the MRC.[65]

As highlighted above, the atmosphere and orientation of the MRC appears to fluctuate according to troubling issues and the political situation in the member countries, or the MRC leadership itself. This can result in inconsistent policies under different CEOs, and also leads to controversy and difficulties within the organisation.

## Greater Mekong Subregion (GMS) Economic Cooperation

Besides the MRC, the GMS is one of the most well-known cooperation forums in the subregion and is one of the few that includes the six riparian nations. As a result, because the GMS includes all Mekong countries, it is important to put a focus on its changing agenda and current programs in regards to regional development in the Mekong subregion.

### *Membership and Funding*

The GMS was initiated by the ADB and the UN Economic and Social Commission for Asia and the Pacific (UNESCAP) as an economic cooperation in Manila in December 1992. The main focus is on building relations based on shared histories and cultures, and the livelihood improvement of the people in the GMS countries. The GMS is a natural economic area connected by the Mekong River and covers approximately 2.6 million square kilometres with a combined population of an estimated 326 million.[66] The nine priority sectors include agriculture, energy, environment, human resource development, investment, telecommunications, tourism, transport infrastructure and transport and trade facilitation. China was represented by the Yunnan Province in the GMS. However, in recent years, China has shown increasing interest in the cooperative development and pushed for leadership, resulting in the additional inclusion of the Guangxi Zhuang Autonomous Region in the GMS.

At the 8th GMS Ministerial Meeting, held in Manila in 1998, the GMS adopted the economic corridor approach. This approach emphasises the development potential of specific geographic areas when infrastructure is planned and developed. The base for these economic corridors are the cross-border transport links that open up opportunities for various types of investments to the otherwise remote and inaccessible areas along the Mekong River.[67] In short, the economic corridor approach links infrastructure and other economic opportunities with development. It includes trade and investment, and efforts to address social and other impacts arising from increased connectivity. The GMS is divided into nine corridors, which transform from transport corridors into economic corridors and are supported through the Economic Corridors Forum, improving interactions between the public and private sector and between local and central governments.[68]

While the GMS directly involves the six riparian countries, many other countries – such as Japan and Australia – have influence on the agenda and programs through their membership in the ADB. The ADB was the founder of the GMS and still plays an important role in its developments today.[69] The influence of the ADB can be seen in different areas including financial, technical and advisory support, secretariat and coordination, and as a broker and catalyst.

In 2014, the ADB received around 65 per cent of its funding from regional countries and approximately 35 per cent from non-regional players (with major non-regional shares held by the US, Germany and Canada). The largest overall shareholder in 2014 was Japan, with a voting power of nearly 13 per cent, followed by China, India, Australia and Indonesia, respectively. The six Mekong riparian countries had a voting power together of nearly 9 per cent in the ADB. China is the second largest shareholder in general and the biggest shareholder out of the Mekong Subregion with just over 5 per cent, followed by Thailand with 1.4 per cent; all other subregional countries holding less than 1 per cent voting power.[70] Further funding bodies within the subregion include the World Bank, the private sector and other governmental and non-governmental sources including the Mekong Fund, The Mekong Development Bank, Thailand's Indochina Fund and the Mekong Development Research Network, as well as several UN programs, such as the UNDP or UNESCAP.

### *Projects and Programs*

The GMS process of institution building can be divided into four phases: first, the confidence-building period from 1991 to 1994; second, the establishment of the framework of cooperation between 1994 and 1996; third, the phase focusing on project preparation; and fourth, the full implementation of high-priority projects since 2000.[71] This shows the transformation of the GMS starting with trust building towards implementing the major plans and projects since 2000. Some projects are still in progress and implementation is often a major burden in the region. As highlighted during the second GMS summit meeting in 2005, important and beneficial projects include the East–West and North–South Economic

Corridors; Regional Power Interconnection and Power Trade Agreements; Flood Control and Water Resource Management; and Tourism Development.

Regional power interconnection and trade is one of the key GMS programs, as about 32 per cent of all energy generated in the GMS is based on hydropower, which is directly linked to the challenges of energy and water security central to this research. Through this cooperation, the benefits for the market-based countries are enormous; however, one of the biggest challenges are the different levels of development and the relative lack of political stability and continuity in some of the member countries, which can slow down opportunities.

Oliver Hensengerth argues that the GMS takes on the economic development role of the earlier MC without a focus on environmental protection. He further argues that as a result of this economic focus, the cooperation is more successful for the region's countries.[72] This book agrees that the GMS focuses mainly on economic development and is currently more successful in regional development. However, it is important to acknowledge the GMS's increasing inclusion of sustainable development and environmental protection. While the GMS's main concerns are in the key areas of energy, forestry, fishing, agriculture and transportation, its ten-year framework proposed in 2001 emphasised a more holistic approach, including the social and environmental implications of infrastructure projects, such as hydropower developments, as well as water resource management to deal with flooding. The ten-year framework was followed by the Midterm Review Report of the Strategic Framework in 2007 and is supported by the Vientiane Plan of Action (VPOA), which was implemented in 2008 and monitors the framework. Further, the GMS shows a growing interest in environmental issues. As stated in a GMS overview report:

> The GMS Program will continue its physical infrastructure investments, but will increasingly focus on softer aspects of subregional cooperation—including human resources, development, trade and transport facilitation, protection of shared environmental resources, and communicable disease control and monitoring [. . .].[73]

Since 2005, the GMS countries have engaged in regional environmental cooperation, launching the Core Environment Program (CEP) and Biodiversity Conservation Corridors Initiative (BCI) in 2006. These were facilitated by the ADB and supported by the Finnish, Swedish and Dutch governments combined with the Poverty Reduction Cooperation Fund from China.[74] The CEP fosters the inclusion of environmental considerations in the project areas of transport, tourism, energy and agriculture and is implemented by the Environment Operations Centre, which opened in Bangkok in 2006.[75] The Centre works in partnership with the GMS governments, local communities, NGOs and research networks to achieve its goals.[76]

The BCI combines biodiversity conservation with poverty reduction and aims at preventing ecosystem fragmentation. The first phase of the cooperation will finish in 2011 and was followed by a strategic framework for the follow-on phase from 2012 to 2016. The CEP–BCI is seen as a major initiative under the GMS Economic Cooperation Program and its importance is highlighted by the ADB. The General Director of ADB's Southeast Asia Department, Kunio Senga, stated that:

> The GMS is one of the most economically robust subregions of the world with most countries posting impressive growth rates exceeding 5–10 per cent in the past two decades. Much of this growth has depended on harnessing the subregion's valuable natural resources. The challenge ahead is how to balance inclusive growth with the sustainable use of natural resources.[77]

This shows the increasing acceptance and inclusion of social and environmental matters into the development of large infrastructure projects, such as dams. However, how effective this initiative will be and whether it can influence investors and beneficiaries of large development projects remains to be seen.

Overall, the principal purpose of the GMS is resource development and infrastructure, not the political stability and security of the Mekong region. However, I argue that the GMS today plays a pivotal role in achieving regional development in the Mekong as well as having an impact on wider Asia–Pacific regional cooperation. The GMS plays an increasingly important role in achieving a balance between environmental degradation and economic development, as both aspects have been included in its agenda and programs since 2001. Further, the GMS includes all six riparian countries as members, and two thirds of its funding through the ABD comes from regional countries, which fosters its acceptance in the region. Supporting this, Vietnamese government advisors and researchers in 2004 found that from a Vietnamese perspective, the MRC is less significant than the GMS.[78] Further, Vietnamese legislation rarely mentions the MRC, although they have a range of national laws and organisations in the area of water.[79] This also refers to local ownership, as until recently 90 per cent of the funding of the MRC comes from foreign donors, showing a lack of interest by the national governments, whereas the GMS is mostly financed by the ADB and Japan.

## Other Subregional Multilateral Institutions

Besides the MRC and the GMS, Thailand initiated the QEC[80] in 1992, which focused on building a strong economy and peaceful ties between the countries based on the concept of 'turning battlefields into market places', linking Thailand, Myanmar, Laos and the Chinese Yunnan Province in the areas of infrastructure and tourism development. The AMBDC is another institutional arrangement in the Mekong Basin that comprises the six riparian countries with China's membership, including the Yunnan Province and Guangxi Zhuang Autonomous Region. Further, the AMBDC is one of the key outcomes of the ASEAN cooperation and was initiated by Malaysia and Singapore in 1996 to enhance economic and sustainable development of the Mekong Basin and to promote the well-being of the people of the Mekong subregion. Its main focus is to create a dialogue and identify common projects that enhance the interconnections and economic links between the riparian countries and the other ASEAN members. The seven main cooperation areas include agriculture; minerals and forestry; industry; transportation; telecommunications and energy; education and training; tourism; and trade and investment.[81]

Another multilateral institution includes the FCDI coordinated by Japan. It focuses mainly on tourism and infrastructure development in CLVM. Moreover, the ASEAN Economic Ministers–Ministry for International Trade and Industry (AEM–METI) initiative combines the ASEAN Economic Ministers Meeting with the Japanese Ministry of Trade and Industry. It focuses on economic growth in Asia as a whole and the economic link between ASEAN and Indochina, especially regarding market transformation, infrastructure development, investment and trade.[82] Further initiatives include the Mekong Fund, Mekong Development Bank, Thailand's Indochina Fund and the International Development and Research Centre-initiated Mekong Development Research Network. The most recent initiative includes the Lancang–Mekong Cooperation Mechanism, which includes all six riparian countries and met for the first time at the foreign minister level in 2015 in the Yunnan Province in China.

These additional institutions and funds are important sources of support for the Mekong subregion; however, the MRC and GMS are the key institutions for regional and sustainable development. A common problem for all regional institutions is that they often cover overlapping memberships and agendas, and show only limited cooperation and partnership between each other. Often, other institutions are seen more as rivals than partners. After highlighting the regional framework of ASAEN, the MRC and the GMS, the third level of analysis focuses now on the micro-level case studies: the Xayaburi Dam, the LS II Dam and the Upper Mekong Navigation.

## Xayaburi Dam – First Mainstream Lower Mekong Dam

The Xayaburi Dam along other hydropower dams is currently one of the most pressing and current issue in Laos related to the Mekong; it receives international media attention and is followed intensively by regional governments, activists, NGOs, development agencies and international organisations. This proposed dam, located in the northern region of Laos, will be the first dam on the Mekong mainstream on the Lower Mekong River (see Map 3.2).[83]

The dam was first proposed by the Lao government in 2010 and the groundbreaking opening ceremony was held at the end of 2012, despite neighbouring countries Cambodia and Vietnam raising concerns regarding possible negative impacts. The Xayaburi Dam is an interesting example because it highlights the difficult situation of the MRC and the PNPCA process. Further, it is also interesting to see how the decision was made in Laos, as the Xayaburi and Nam Theun 2 dams are the only two dams authorised by the National Assembly

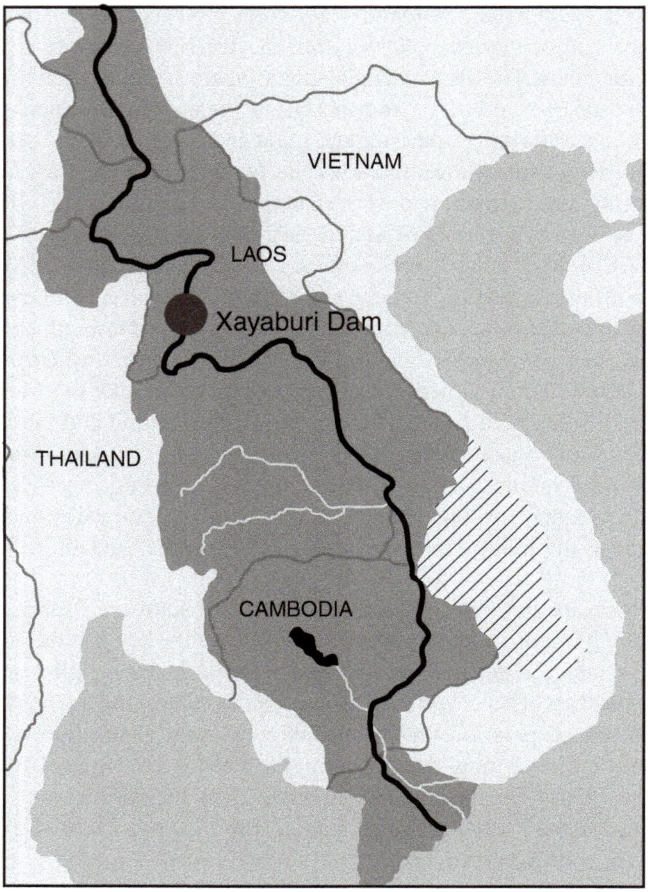


*Map 3.2* Xayaburi Dam in Laos.

Source: Adjusted from Oxfam Australia, 2013.

(NA) of Laos. The Xayaburi Dam – currently being built in the Laos province of Xayaburi – will stretch along the whole Mekong River. The project is estimated to produce around 1,200 megawatts of electricity, of which around 95 per cent will go to Thailand.[84] It is estimated that it will take around eight years to finish with an estimated total cost of 3.6 billion USD.[85] According to the Lao government, the Xayaburi hydropower dam is expected to bring 130 million USD a year in royalties and taxes.[86]

Key developers of the project include Thailand's Ch. Karnchang Public Company, the Electricity Generating Authority of Thailand (EGAT),[87] the Lao government and Ratchaburi Electricity Generating Holding Company.

A Memorandum of Understanding with Thailand's Ch. Karnchang Company as developer was signed in late 2008 and was followed by a Power Purchasing Agreement with EGAT and the Lao government in mid-2010, allowing EGAT to purchase electricity at a cost of 2.159 Thai Baht per kWh transferred through a 200-kilometres-long transmission line into northern Thailand.[88]

Because Thailand provides the market for the produced electricity, Thailand's Ministry of Energy and Electricity and EGAT play key roles in the developments. This also includes Ch. Karnchang, which is Thailand's second largest publicly traded construction company and, therefore, has an impact on the country. As Thailand has already developed most of its domestic hydropower potential, it seeks to import from the neighbouring countries to secure its growing energy demand. Also, previous strong opposition to dam projects including the Pak Mun Dam further supports the increasing involvement in hydropower development abroad, mainly in Laos, where civil society is weaker. Often, the Xayaburi Dam is seen as a Thai project, since Thailand provides the financing, the labour force, the construction and buys the electricity.

Estimated impacts of the Xayaburi Dam include the loss of biodiversity, a change in the ecosystem and the reduction of fish. A highly controversial point is the use of fish ladders on the Xayaburi Dam, as proposed in the first draft of the dam. Fish experts argue that these ladders would not work due to the high diversity of fish in the Mekong, including different sizes and high number of fish, and the height of the dam. It is estimated that around 2,100 people would be resettled by the project and that more than 202,000 people living on the river banks in close proximity to the dam site would be affected through loss of income and food security, including loss of agricultural land, loss of opportunities for gold panning[89] and lack of access to forest products.[90]

Besides the impact on biodiversity, fish stocks and food security, another problem of the proposed dam is that the actual cost of the already existing revenues from a 'healthy' Mekong River is not counted and social and environmental damages, as well as the necessary money for mitigation and compensation, are missing from any calculations. This means that the actual cost is unknown and cannot be weighed against the income generated through the dam investment. Another key issue often discussed is that the costs and benefits will be unevenly distributed between the people within Laos, but also between the riparian countries in the Lower Mekong.

### *Procedures for Notification, Prior Consultation and Agreement (PNPCA) and the Role of the MRC – Any Role?*

In September 2010, the Lao government submitted the proposal of the Xayaburi Dam to the other regional countries through the PNPCA of the MRC. This process is based on Article 5 of the 1995 Mekong River Agreement, which defines 'reasonable and equitable utilisation' of the river.[91] Section B 2a of the same article requires that countries notify, consult and reach agreement with the member riparian states if they utilise the waters on the mainstream of the

Mekong River for intra-basin use during the dry season, which is the case in the Xayaburi Dam.

The PNPCA process of the MRC was used on the Xayaburi Dam for the first time and was seen as a test. At the same time, the MRC released a major Strategic Environment Assessment (SEA) report stating that mainstream dams on the Mekong should be deferred for another ten years due to massive possible risks and unknowns impacts on the river,[92] which further demonstrates the difficult position of the MRC. The Xayaburi Dam is going ahead despite concerns of the other riparian countries, Cambodia and Vietnam, regarding the PNPCA. One expert stated that 'within its mandate, the MRC handled the PNPCA very well. The MRC didn't do anything wrong but the PNPCA ruined the reputation of the MRC'.[93]

While Cambodia and Vietnam were outspoken against the project until the end of 2012, since then the rhetoric from both countries has waned. The reasons for this are unclear and different opinions exist as to why Vietnam reduced its pressure on the Lao government. Some experts argue that the redesign of the Xayaburi Dam was shared bilaterally with Vietnam – although the MRC itself did not receive the documents until July 2013. Others argue that a bilateral deal between both countries was achieved, which could include other areas such as rubber plantation or logging along the border between Laos and Vietnam or the building of the Luang Prabang Dam by a Vietnamese developer. In the Vietnamese media, pressure against hydropower development still exists; however, it does not specify the Xayaburi Dam anymore. Further, Vietnam progresses with its studies in the Mekong delta focusing on impacts from upstream dams. Overall, Vietnam does not have a strong standpoint against dam building because around 50 per cent of the Mekong river flow comes from the Vietnamese Highlands, which got blocked by dams built in Vietnam.[94]

The Lao government perspective – which Thailand supports – is that the engineers altered the Xayaburi Dam's design to focus on the fish passages, which meant that the project could proceed. Since international pressure focused on fish passages, as opposed to resettlement or water flow, it was easy for the government to adjust this one area. The Xayaburi Dam and the Nam Theun 2 Dam are the only two dams authorised by the NA of Laos. Experts argue that the Xayaburi Dam was authorised by the NA to shift the focus away from the Ministry for Energy and Mines and the opposing views within the Lao government.[95] Further, the Lao government aims to overcome the country's situation of being one of the least developed in the world, and hopes to invigorate economic development and move from being a landlocked country to a land-linked one.

As stated by the Lao Vice-Minister for Energy and Mines, Dr Viraphonh Viravong, hydropower is important because it is clean, cheap and renewable, further affirming that 'hydropower contributes something like 33 per cent to the natural capital of the wealth of Laos. And if Laos wants to leave behind its least developed country status by 2020, this is our only choice'.[96] Moreover, decision-makers in the country will benefit sufficiently from large construction projects. Similarly, large constructions are still seen as a sign of power and a symbol of a growing and influential Laos.

In this context, it is also important to remember that the projects that proceed in the Mekong region are often not only due to technical but also political decisions. Similarly, government departments in the countries do not have one position; often, there are several drivers behind the decision-making process. Further, controversies regarding the Xayaburi Dam include the low fixed price Thailand pays for electricity generated from it. The price is already relatively low by today's standards; however, this anomaly will be even more prominent over time when energy prices increase.[97] The Finnish engineering company Pöyry's[98] role in the process was also an unexpected move by the Lao government. The company carried out the Environmental Impact Assessment (EIA) and then became the lead engineer of the project.[99] In a similar manner to other hydropower developments in the region, this provides a 'conflict of interests', which means that a company has an advantage by making certain statements and approving projects independently of possible social and environmental impacts.[100] In general, hydropower concentrates wealth and distributes negative impacts; banks make profits and how the government uses the income through revenues is often unclear.[101]

### *Conclusion: Xayaburi Dam*

The Xayaburi case shows the problematic nature of large infrastructure developments in the region and their impact on the local population and the uneven distribution of wealth, especially among the poorer rural communities. As the first mainstream dam on the Lower Mekong River, the Xayaburi Dam will surely herald future mainstream dam projects since a precedent will have been set. Future developers will argue that the first dam has the biggest impact and that if the Xayaburi Dam goes ahead, other dams are fine as well. This seems the likely development on the Mekong River as there are several other proposed mainstream dams, one being the Don Sahong Dam in southern Laos in close proximity to the Cambodian border.[102] However, this view would neglect the problem of cumulative impacts on the river and sediment flow through several dam projects, as these studies are often not appropriately included in single projects, which ignore the problem of transboundary effects downstream. Furthermore, the Xayaburi case is a unique case as the PNPCA was used for the first time. However, most would argue that this was unsuccessful as the dam construction went ahead, although the MRC recommended further studies on the impacts of mainstream dams. Similarly, the downstream countries Cambodia and Vietnam still do not support the project and maintain that the PNPCA process was not finished.

## Lower Sesan II (LS II) – Just a Tributary Dam?

The LS II Dam is not a mainstream dam like the Xayaburi Dam; rather, it is a tributary dam to the Mekong River. In some respects, this makes this case different as the focus is often mainly on mainstream dams and their impact on the rest of the river basin. However, as argued by several scholars and scientists, the negative effects of hydropower development on the Sesan River such as the LS II can be enormous and would not only have a massive impact on Cambodia itself, but also on Vietnam.[103] The impacts could be felt as far downstream as the Mekong Delta and as far upstream as Laos and Thailand. Being a tributary dam also has major implications in regards to the involvement of the MRC and the PNPCA. In the past, Cambodia had less money, capacity and investors in comparison to Laos. However, today the levels are similar due to China's increasing interest in Cambodia. Whereas Laos wants to use hydropower mainly for export, Cambodia wants to use electricity also for the domestic market, to build the grid and to reduce the currently high energy prices in the country.[104] A critical factor in Laotian and Cambodian hydropower decision-making is that dams represent national symbols: having a large visible dam shows progress.[105]

The Sesan River is one of the largest Mekong River tributaries. It originates in the Vietnamese Highlands and flows through Cambodia before ending in the mainstream of the Mekong (see Map 3.3[106]). Together, the Sesan, Sre Pok and Sekong Rivers[107] add around 10.4 per cent of the total flow of the Mekong.[108] The LS II Dam is located near the joining point of the Sesan and Srepok rivers in Stung Treng Province in north-eastern Cambodia and will threaten the biodiversity and fishery of two of the Mekong River's most significant tributaries while also dislocating thousands of people.[109]

The basic features of the LS II include a capacity of 400 megawatts, a height of 75 metres, a length of six kilometres and a reservoir size of 33,560 hectares, with an estimated cost of 781 million USD.[110] The LS II project was first envisaged in April 2011 by the Cambodia–Vietnam Hydropower Company, a joint

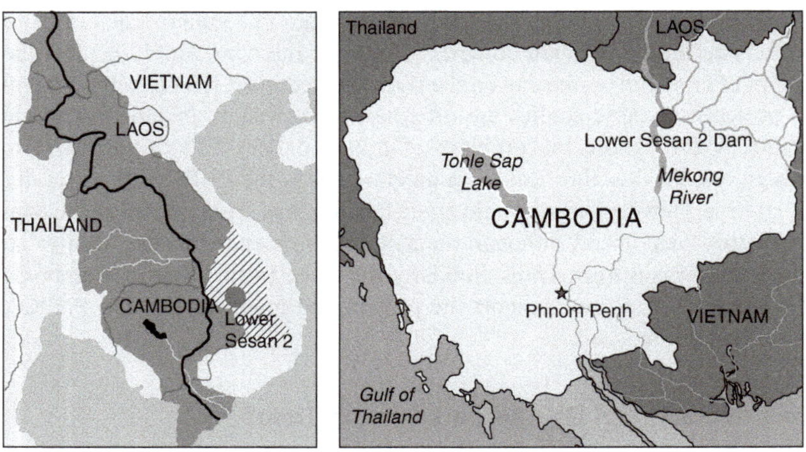


*Map 3.3* Lower Sesan II Dam location in Cambodia.

Source: Adjusted from Oxfam Australia, 2013; Radio Free Asia, 2012.

venture between the developers (the Electricity of Vietnam International Joint Stock Company (EVNI)), which held 51 per cent, and the Royal Group, which held 49 per cent.[111] EVNI is a parastatal, semi-independent government agency and the Royal Group is owned by one of the largest mobile networks in Cambodia and a 'favoured son' of Hun Sen, the Prime Minister of Cambodia. However, this relationship did not hold, and in November 2012 China's Hydrolancang International Energy, a large Chinese international company, took over and signed a joint agreement with the Royal Group. It is unclear whether EVNI might still be involved in regards to building the transmission lines.[112]

Overall, it is estimated that the project will take five years to complete and expected revenues for the dam are 29.5 million USD per year in tax once electricity generation begins.[113] The Cambodian government approved the project at the end of 2012 despite the EIA report assessing that it failed to meet international best practice. The financial plan of the LS II was accepted by Cambodia's NA in February 2013, construction started in February 2014 and the first resettlement followed in mid-2015.[114]

The ethical implementation of the LS II has been questioned since Hydrolancang recently violated Chinese law when building a dam passage on the Upper Mekong in China.[115] Also, the economic viability of the project has been questioned due to an increase in drought fostered by climate change and changes to the water flow resulting from upstream dam operations (as the Sesan River is already heavily dammed in Vietnam). Ecologically, the planned project could have devastating impacts as scientists found that the LS II could cause a 9.3 per cent drop of fish stock basin-wide and threaten more than 50 fish species.[116]

Further, the LS II Dam will contribute to the changing of the Mekong's hydrological flows, as well as the water levels of the Mekong downstream of the Tonle Sap Lake, including severe reduction of sediment flows. Resettlement is complicated in the LS II area as a large number of people (approximately 5,000) need to be relocated. This is especially difficult as many are from ethnic minorities with a special claim on land through cultural ties.[117] This will influence the livelihoods and cultures of tens of thousands of people living along the Sesan and Srepok Rivers, who are closely connected to the river system and its rich natural resources. Affected people have yet to be properly consulted and resettlement plans have not been disclosed to the public.[118]

The civil society and NGO networks active in Cambodia such as the River Coalition of Cambodia (RCC) or the NGO Forum (NGOF) advocate their standpoint and influence the Cambodian government. However, sometimes this influence is very limited and weak as the RCC members cannot achieve consensus (e.g. some members want to oppose the construction of the LS II while other members do not oppose it but instead want to secure a fair and decent compensation and proper relocation arrangements).[119] These different views make it difficult to voice one opinion, which is important in order to achieve anything in a context where civil society already has a difficult standpoint and public participation is often limited or non-existent.

### *The Role of the MRC*

As highlighted earlier, Article 5 of the 1995 Mekong River Agreement defines the reasonable and equitable utilisation of water between tributaries and the mainstream. Section A states that 'on tributaries of the Mekong River, including Tonle Sap, intra-basin uses and inter-basin diversions shall be subject to notification to the Joint Committee'.[120] This means that tributary dams are within national jurisdiction and countries need to 'notify' the MRC rather than consult it.[121]

However, experts argue that the 1995 MRC has a basin-wide approach despite Articles 5–8 of the agreement specifically referring to mainstream use. Experts argue that these articles are related to the effects on the mainstream but not if the project itself is located on the mainstream or a tributary. This is especially important as Ziv *et al.* argue that 'dams at different locations have different trade-offs between power generation and the loss in fish biodiversity and productivity', with the LS II Dam having 'the highest environmental cost per unit of energy produced'.[122] As a result, due to the considerable future impacts associated with the LS II hydropower dam, the project should have been notified to the MRC and consultation with other MRC member states should have been conducted prior to approval.

However, the MRC itself cannot do much about the LS II without being notified (which is the current status in 2015). While it has not done so yet, Vietnam could call upon the MRC by referring to Articles 7 and 8 of the MRC 1995 agreement, which focus on the prevention and cessation of harmful effects whereby states can notify other states with proper and valid evidence that shows 'substantial damage to one or more riparians from the use of and/or discharge to water of the Mekong River'.[123] Although this could be an option for Vietnam, it seems unlikely because of Vietnam's own projects in the 3 S Rivers, which resulted in previous negative impacts on the subregion, mainly in Cambodia. In regards to effects from the LS II on Cambodia itself, the MRC cannot get involved as this is a national issue. Similarly, as Cambodia has never informed the MRC about the project or shared any documents on the project with the MRC, the organisation is without a mandate and does not comment on it.

As a result, the MRC's involvement seems unlikely because no party will ask for support and it cannot act without a mandate by a riparian state.[124] International donors of the MRC urged the Cambodian government in June 2013 and early 2014 to redesign the LS II project and submit it to the MRC's prior consultation process.[125] However, although the 'LS II should be submitted to the PNPCA', it has not been because Cambodia has not submitted it yet.[126]

### *Conclusion: LS II*

The LS II Dam shows that tributary dams can be as significant as mainstream dams and should not be neglected. This case study highlighted that the influence and impact of the LS II could possibly be bigger than the Xayaburi Dam due to its location impacting on water flow, fish stocks and local communities, including the large number of resettlements necessary for the project. This case shows again the lack of a strong MRC agreement and the 'tied hands' of the MRC regarding the LS II Dam project and other current dam developments along the Lower Mekong. It is unlikely that the MRC will become involved in this project in the future as it does not have a mandate. Also Vietnam's past dam building on the 3 S Rivers and the still possible involvement of EVNI suggest that Vietnam will not call upon the MRC.

## Upper Mekong Navigation – Rapid Blasting

Besides hydropower projects, navigation also plays a role in the Upper Mekong, especially between Simao in China and Luang Prabang in Laos (see Map 3.4).

Since the 1980s, China has played an increasing role in the Mekong River developments under its 'Go West' policy and the Lancang[127] Economic Belt – comprising hydropower dams, industrial estates and navigation channel improvement projects for large cargo ships on the Upper Mekong.[128] China has been dredging its part of the Lancang since the 1960s and has made a nearly 300 kilometres stretch navigable for 100–300-ton vessels.[129] In 1992, China initiated a plan for waterway alteration to open up the northern section of the river to all-year-round navigation for large cargo ships. The navigation project goal was to make the 886 kilometres stretch from Simao in China to Luang Prabang in Laos navigable for vessels up to 500 tons.[130]

This improved navigation project on the Upper Mekong started with the Lancang–Mekong Navigation Channel Improvement Project in 2000 funded by China as a component of the GMS Program focusing on a growth triangle in the subregion. The goal was to open river navigation to commercial vessels and provide an access route between the fast-growing economy of China and the Lower Mekong, especially Thailand.[131] Laos, Thailand and Myanmar formally agreed to the project and signed a navigation agreement in 2000, which would allow the participating countries to trade freely along the Mekong River for 886 kilometres.[132] In March 2002, rapid blasting along the Mekong started and China provided 5 million USD to finance the project.[133] In the first phase, the project destroyed 11 major rapids and ten scattered reefs along the 331-kilometres-long section on the Mekong before channelling the river in the second and third phase of the project.[134]

Impacts on the environment and livelihoods are real, as rapids provide a critical habitat for hundreds of migratory fish species when they move upstream to feed and breed. However, rapids impede navigation and need to be destroyed, thus affecting flora and fauna in the river. For instance, the rapids and rocks near Chiang Khong in Thailand are the only known spawning grounds for the endangered Mekong catfish and are affected by the blasting.[135] Additionally, reefs also play a key role in producing oxygen, reducing pollution and aiding the decomposition of vegetation within a river basin such as the Mekong.[136] Local agriculture could also be affected through the changes of river current

*Map 3.4* Navigational route from Simao to Luang Prabang.

Source: Adjusted from Mastny, 2003: 23.

affecting the river banks and, therefore, important agricultural land along the river banks. Local people discovered increased erosion, higher sediment levels and strange disruptions of the water flow.

Further, an increase in commercial navigation – especially the shipping of dangerous goods – could have a devastating impact on the environment and livelihoods within Laos, especially in the northern part. Several villages found unusual water fluctuation patterns, an intensification of river bank erosion and a decrease in sediment load impacting on the ecological function of the Mekong.[137] Besides the impact on tropical freshwater fish and rising water levels, the changes will also impact the local villages collecting Mekong water weed, which is an important source of food and protein and also a well-known delicacy from the world-heritage city Luang Prabang. The unusual rapid water fluctuation also affects the rich riverine ecosystems, local livelihoods and fishing communities; for instance, on the Thai–Lao border, fish catch declined by 50 per cent (between the start of the blasting and 2004) due to water fluctuation.[138] Furthermore, the increased presence of large vessels presents hazards for small fishing boats, which can be affected directly or through the waves on the river. In the long run, the water flow to the sea will be faster due to a decrease in resistance, which will lead to an increased risk of droughts and floods, including flash floods. More pollution due to an increased navigation upstream might provide bigger impacts on the environment and long-term livelihoods.

China's movement in the Golden Triangle shows the importance of the trade route for the future; China wants clear and open navigational rights on the Upper Mekong.[139] Laos is strategically located in regards to the economic developments in the upper region; however, the country seems to be more a passage rather than a destination and actually benefiting from the navigation. 'In most cases, the poorest river dwellers – most of them in Laos – lack the education and skills to benefit from these new opportunities. For them, the river is still sustenance and changes to it pose a dire threat'.[140]

Additionally, the navigation channel needs continuous maintenance and extensive dredging to keep the Mekong navigable. This will be linked to future cost for the riparian countries such as Laos. However, it is also important to keep in mind that there were large investments from China in Laos, including direct payments for the financing of the blasting, but also for other development projects in Laos, which again show the increasing impact of China in the Lower Mekong countries and the interconnectedness of the region.

In general, the impact of the navigation project will be felt more locally than large hydropower projects. The improved navigation is beneficial for China and Thailand, but Thailand also has a loss of revenue through cheap imports from China and an influx of Chinese business within its borders. Incoming Chinese vessels flood the local markets with cheap products including textiles and other merchandise, which are up to five times cheaper than domestic goods, threatening local shopkeepers and farmers along the Mekong in Thailand and Laos.[141] This view is also supported by a Thai official in Chiang Rai Province who argues that the majority of cargo ships entering the Chiang Saen Port in 2001 were Chinese; of the 2,400 ships entering, more than three-quarters originated in China.[142]

While the navigation project will directly impact Thailand, Myanmar, Laos and China, it will also have an impact on people in downstream Cambodia and Vietnam. Officials have raised concerns that the project could alter water flow, cause riverbank erosion and increase pollution when accidents occur. However, the downstream countries Cambodia and Vietnam have not been consulted and in Thailand, unprecedented protests were held in 2002 when more than 12,000 people rallied against the blasting and navigation projects.[143]

### *The Role of the MRC – Any Impact?*

An agreement between the four riparian countries on Upper Mekong Navigation was signed in 2000 and the Joint Committee on Coordination of Commercial Navigation (JCCCN) was formed. As well as hydropower development, navigation is also mentioned in the areas of cooperation in the 1995 MRC agreement, and recently the MRC has become more involved in this area.[144] As stated by a high-level MRC expert, commercial navigation on the Upper Mekong is nowadays more of an issue due to an increase in navigation, but also because of the possible negative effects of it.[145] Environmental impacts and consequences can occur due to an increase of navigation but also due to the transportation of dangerous goods such as liquefied petroleum gas, oil and other petroleum products from Thailand into China.[146] The former MRC CEO Joern Kristensen stopped the blasting in 2002.[147] Further, some rapid blasting along the Thai–Lao border was stopped in 2003 because of demarcation problems and concerns about changing the border between both countries.[148]

Overall, the Lao or Thai government did not approach the MRC in relation to navigational issues associated with this project (that was in-line with the MRC guideline of freedom of navigation). However, risks such as environmental spills are still present. An EIA was sponsored by the Chinese in September 2001 stating that the project impact will be negligible, only affecting the area directly surrounding the blasting itself.[149] However, the EIA was criticised by many and the MRC sponsored its own expert review of the EIA and supported the Lao government in the impact assessment. The three assessments funded by the MRC focused on the impacts on fisheries, livelihoods and socio-economic aspects.[150] All three reports highlighted the problems and insufficient research of the Chinese EIA and stated that more research needed to be done to fully understand the possible impact of the rapid blasting and other effects related to the goal of achieving an improved navigation channel. This shows the validity of Tyson Roberts's statement regarding EIAs that,

> social and environmental assessment is valid only if undertaken on behalf of rural as well as urban society and on behalf of the environment. Assessments done by or controlled by organisations promoting the projects are seldom reliable. Such assessments usually overestimate economic and other benefits while ignoring or underestimating social and environmental impacts.[151]

This symptom is similar to the conflict of interest highlighted earlier regarding Pöyry's role in the Xayaburi Dam development.

### *Conclusion: Upper Mekong Navigation*

Navigation in the Upper Mekong, besides hydropower developments, has significant implications for the Mekong subregion. The MRC is currently not closely involved in the issue of navigation but a bigger engagement is possible in the future. Most likely this will be in the areas of handling dangerous goods, improved navigation roads and the installation of water identification marks for safety matters.[152] Navigation is currently not so prominent due to the rapid developments on mainstream dams; however, this can change in the future.[153]

The MRC became involved in the assessment of the impacts of rapid blasting on behalf of the Lao government. In the future, involvement could be due to an increase of commercial navigation and its possible negative impacts through the transportation of dangerous goods. Similarly, an increase of environmental pollution through navigation in the future is likely. The impact will be more local – in comparison to large hydropower projects – but still visible. Navigation also has an impact on trade and commerce in positive and negative ways, with winners and losers as highlighted above. China seems to be cooperative and active in achieving the Upper Mekong Improvement project. However, as highlighted above, China was and is the main beneficiary due to an increase in trade with the other Lower Mekong countries, Chinese labour and goods spreading actively in the region and the progress of its Yunnan Province.

## Summary and Conclusion

Hydropower developments on the mainstream and tributaries of the Mekong besides other large infrastructure projects are the current challenges in the subregion affecting livelihoods along the river. Navigation could also become a more contested topic in the future due to more pollution as a result of improved navigation and an amplified risk of accidents impacting on the environment.

This chapter has argued that decision-making in the Mekong River Basin is interlinked with the political agenda and other interests of the riparian states. Decision-makers and politicians involved and their linkages are often the drivers behind decision-making regarding large infrastructure developments. As previously highlighted, the role of the MRC is limited by a very restricted mandate which makes it difficult for it to be involved in dam developments such as the Xayaburi Dam or the Lower Sesan II. The situation is similar in regards to navigational matters, where the MRC has been less involved in the past. The PNPCA process and its first-time use further weakens the role of the institution, as the MRC was unable to influence the riparians according to their findings and recommendations regarding dam developments on the mainstream of the Mekong. However, these developments could be a dangerous starting point, as it might become increasingly difficult to stop future projects which could, in the long run, also impact on navigational interests.

The Mekong case study highlighted the importance of the Mekong subregion for the economic progress of the individual countries but also for regional cooperation. It further focused on the need to find a balance between economic growth and the protection of the natural resources, including the existing diverse ecosystem on which so many people depend. This case also provided an insight into developments on the Mekong River, which includes problems and suspicions between the Mekong countries because of the ideological divide during the regional wars in Vietnam, Laos and Cambodia, and can be seen even today in small-scale conflicts regarding temple sites, border demarcation or islands in the Mekong. The Mekong case study further showed that ASEAN is an umbrella for the subregion, yet environmental matters and guidelines do not influence the subregion as strongly as it is the case, for instance, on the Danube River with the binding EU Water Framework Directive. In the ASEAN context, non-interference and non-binding agreements still prevail and, therefore, impact on the member countries differently, with one of the key challenges being the implementation from regional frameworks into national policies and law.

Lastly, this case stressed the importance of subregional institutions such as the MRC and the GMS and highlighted their membership, funding and programs for the Mekong. It also provided insight into the problems of cooperation between the existing institutions with often overlapping agendas and members. Moreover, the chapter showed the rising importance of the GMS program and its increasing inclusion of social and environmental matters in large development projects. The GMS has developed, in the last decade, a focus on environmental issues and can be the driver for future successful cooperation in the Mekong subregion. The MRC is an important body regarding research and data collection on water resource management. However, it is limited in its mandate, which often weakens the riparian body and limits its success.

## Notes

1 This ranking is controversial and depends on the measuring of the river by different methods. Different methods include measuring by the mean annual discharge or the length of the river. More details about the methods and problems can be found in Campbell (2009).

2 Milton Osborne, "The Strategic Significance of the Mekong," *Contemporary Southeast Asia* 22, no. 3 (2000): 430; Evelyn Goh, "China in the Mekong River Basin: The Regional Security Implications of Resource Development on the Lancang Jiang," in *Non-Traditional Security in Asia*, eds. Mely Caballero-Anthony, Ralf Emmers, and Amitav Acharya (Hampshire and Burlington: Ashgate, 2006), 230.

3 Including the area of present-day Cambodia and parts of today's Thailand.

4 Charles Higham, *The Civilization of Angkor* (London: Weidenfeld and Nicholson, 2001).

5 Myanmar is also known as Burma.

6 Central Intelligence Agency (CIA), "Laos," www.cia.gov/library/publications/the-world-factbook/geos/la.html (accessed 23.11.2015); CIA, "Cambodia," https://www.cia.gov/library/publications/the-world-factbook/geos/cb.html (accessed 23.11.2015).

7 Osborne, "The Strategic Significance of the Mekong."

8 Ian C. Campbell, "Chapter 1: Introduction," in *The Mekong – Biophysical Environment of an International River Basin*, ed. Ian C. Campbell (New York: Elsevier, 2009), 5.

9 Katri Mehtonen, Marko Keskinen, and Olli Varis, "The Mekong: IWRM and Institutions," in *Management of Transboundary River and Lakes*, eds. Olli Varis, Asti K. Biswas, and Cecilia Tortajada (Berlin: Springer, 2008), 211.
10 Milton Osborne, *The Mekong: Turbulent Past, Uncertain Future*, 2nd ed. (Crows Nest, Australia: Allen and Unwin, 2006), 59.
11 Anchalee Kongrut, "China Should Set Good Example on the Mekong River," *Economist Observer*, 25.01.2013.
12 Interview with an expert in water management in the Mekong Basin based in Vientiane, Laos on the 24.05.2013.
13 *Wall Street Journal*, "Tension over Dam Project Shifts Myanmar's Politics," *The Wall Street Journal*, 13.08.2011.
14 CIA, "Laos."
15 Interview with a high-level expert of the MRC in Vientiane, Laos on the 06.06.2013.
16 *Daily Fusion*, "China Becomes 'Key Investor' in Southeast Asia's Hydropower," *Daily Fusion*, 19.08.2013.
17 The Pak Mun Dam on the Mun River in north-eastern Thailand, started in 1991 and completed three years later, was one of the most controversial dam buildings on a Mekong tributary. It was built under the Electricity Generating Authority in Thailand (EGAT) and was intended to provide 136 megawatts of hydropower; however, it failed to reach the target and instead produces around one third of it. Significant social costs included covering the relocation of the population, which is estimated to be seven times higher than previously calculated, as well as the enormous decrease in fish stocks and species reduction (169 out of 265 species disappeared). Osborne, "The Strategic Significance of the Mekong," 440.
18 Interview in Bangkok on the 28.11.2012.
19 Osborne, "The Strategic Significance of the Mekong."
20 So Nam, "Fisheries Resources in Cambodia – Current Status, Key Issues and Directions," Inland Fisheries Research and Development Institute, n.d.
21 David Phillips, Marwa Daoundy, Stephen McCaffrey, Joakim Ojendal, and Anthony Turton, *Trans-Boundary Water Co-Operation as a Tool for Conflict Prevention and Broader Benefit Sharing*, ed. Global Development Studies No. 4. Ministry of Foreign Affairs, Stockholm, 2006, 230.
22 *Daily Fusion*, "China Becomes 'Key Investor' in Southeast Asia's Hydropower."
23 Jon Fernquest, "Rice Exports: Fall from 1st to 3rd," *Bangkok Post*, 25.10.2012.
24 For an overview of proposed and built dams in Vietnam see: International Rivers Network, "Planned Dams in Vietnam," International Rivers Network, 2001.
25 *Daily Fusion*, "China Becomes 'Key Investor' in Southeast Asia's Hydropower."
26 Environmental Investigation Agency, "Crossroads—The Illicit Timber Trade between Laos and Vietnam," Environmental Investigation Agency, London, 2011.
27 Ibid. *Crossroads* provides a detailed and in-depth look into the illicit timber trade between Laos and Vietnam and offers a good insight into the problematic situation in the region, which can be transferred to other areas.
28 The ten ASEAN member states are: Indonesia, Malaysia, the Philippines, Singapore, Thailand, Brunei, Vietnam, Myanmar, Laos and Cambodia.
29 Association of Southeast Asian Nations (ASEAN), "Overview of the Association of Southeast Asian Nations," www.asean.org/asean/about-asean/overview/ (accessed 29.04.2015).
30 Amitav Acharya, *Constructing a Security Community in Southeast Asia: ASEAN and the Problem of Regional Order* (New York: Routledge, 2001), 63. Amitav Acharya, "Ideas, Identity, and Institution Building: From the 'ASEAN Way' to the 'Asia–Pacific Way'?," *The Pacific Review* 10, no. 3 (1997): 329.
31 Ibid., 329.
32 Heribert Dieter, "Changing Patterns in Regional Governance: From Security to Political Economy?," *The Pacific Review* 22, no. 1 (2009): 87.

33 Derek McDougall, "International Relations," in *Understanding Contemporary Asia Pacific*, ed. Katherine P. Kaup (Boulder, CO: Lynne Rienner, 2007), 184; Sheldon W. Simon, "ASEAN and the New Regional Multilateralism," in *International Relations of Asia*, eds. David Shambaugh and Michael Yahuda (Lanham, MD: Rowan & Littlefield Publishers, 2008), 202.
34 Quoted in John Ravenhill, "Fighting Irrelevance: An Economic Community 'with ASEAN Characteristics'," *The Pacific Review* 21, no. 4 (2008): 469.
35 Rodolfo C. Severino, "ASEAN: New Charter, New Optimism," *Southeast Asia Security Outlook* 153, 2009/2010 (2010): 7–9.
36 Quoted in ibid., 9.
37 Kenneth P. Tan, "Crisis, Self-Reflection, and Rebirth in Singapore's National Life Cycle," *Southeast Asian Affairs* (2003): 243.
38 Richard Stubbs, "Meeting the Challenges of Region-Building in ASEAN," in *Contemporary Southeast Asia*, ed. Mark Beeson (Hampshire: Palgrave Macmillan, 2009), 235.
39 Lorraine Elliott, "ASEAN and Environmental Cooperation: Norms, Interests and Identity," *The Pacific Review* 16, no. 1 (2003): 36, 38.
40 Apichai Sunchindah, "The ASEAN Approach to Regional Environmental Management," ASEAN Secretariat, 2002.
41 ASEAN, "Hanoi Plan of Action," www.asean.org/?static_post=hanoi-plan-of-action (accessed 22.05.2016).
42 Section D of the Blueprint deals with 'Ensuring Environmental Sustainability.'
43 ASEAN Cooperation on Environment, "Overview of ASEAN Cooperation on Environment," http://environment.asean.org/about-us-2/ (accessed 30.09.2013).
44 Kheng L. Koh and Nicholas A. Robinson, "Regional Environmental Governance: Examining the Association of Southeast Asian Nations (ASEAN) Model," in *Global Environmental Governance*, eds. Daniel C. Esty and Maria H. Ivanova (New Haven, CT: Yale School of Forestry and Environmental Studies, 2002), 105.
45 Evelyn Goh, "ASEAN Adventures," *The National Interest* 94 (2008): 63.
46 Quoted in Stubbs, "Meeting the Challenges of Region-Building in ASEAN," 242.
47 Osborne, "The Strategic Significance of the Mekong," 433.
48 Ibid.
49 Osborne, "The Strategic Significance of the Mekong."
50 Abigail Makim, "Resources for Security and Stability? The Politics of Regional Cooperation on the Mekong, 1957–2001," *Journal of Environment and Development* 11, no. 1 (2002): 18.
51 The Yali Falls Dam is the second largest dam in Vietnam, located on a tributary of the Sesan River (which flows into the Mekong) and originates in the Central Highlands of Vietnam, about 70 kilometres upstream of Cambodia.
52 Mekong River Commission (MRC), "Agreement on the Cooperation for the Sustainable Development of the Mekong River Basin," MRC, Chiang Rai, Thailand, 1995, Article 1.
53 MRC, "Annual Report 2001 – Mekong River Commission," MRC, Vientiane, 2002; MRC, "About the MRC," www.mrcmekong.org/about-mrc/ (accessed 23.05.2016).
54 Mya Than and George Abonyi, "The Greater Mekong Subregion: Co-Operation in Infrastructure and Finance," in *ASEAN Enlargement: Impacts and Implications*, eds. Mya Than and Carolyn Gates (Singapore: Institute of Southeast Asian Studies, 2001), 132.
55 MRC, "Annual Report 2001 – Mekong River Commission," 2.
56 MRC, "Mekong Prime Ministers Agree to Prioritise Climate Change as Summit Ends," www.mrcmekong.org/news-and-events/news/mekong-prime-ministers-agree-to-prioritise-climate-change-as-summit-ends/ (accessed 23.05.2016).

57 MRC, "MRC Financial Statements and Independent Auditor's Report Year Ended 31 December 2014," MRC, Vientiane, 2015, 1–18.
58 This is part of a larger restructuring the MRC is undergoing as part of the Strategic Plan 2016–2020.
59 Jacob Hook, Susan Novak, and Robyn Johnston, "Social Atlas of the Lower Mekong Basin," MRC, Phnom Penh, 2003, 10.
60 Olivier Cogels, "Mekong Hydropower Development Is Good," *Bangkok Post*, 09.01.2007.
61 Children of Mekong, "Mr Jeremy Bird," www.childrenofthemekong.org/ (accessed 04.11.2011).
62 International Center for Environmental Management, "Environmental Assessment of Hydropower on the Mekong Mainstream – Summary of the Final Report," International Center for Environmental Management, Hanoi, 2010.
63 MRC, "MRC Appoints New Chief Executive Officer," www.mrcmekong.org/news-and-events/news/mrc-appoints-new-chief-executive-officer/ (accessed 05.07.2012).
64 MRC, "Sweden Provides US $8.5 Million to Strengthen Sustainable Development and Management of Mekong Resources," www.mrcmekong.org/news-and-events/news/sweden-provides-us-8-5-million-to-strengthen-sustainable-development-and-management-of-mekong-resources/ (accessed 25.07.2012).
65 This is part of a larger restructuring process the MRC is going through as part of the Strategic Plan 2016–2020.
66 Greater Mekong Subregion (GMS) Secretariat, "Greater Mekong Subregion," www.adb.org/gms (accessed 17.10.2011).
67 Ibid.
68 Ibid.
69 Mehtonen *et al.*, "The Mekong: IWRM and Institutions."
70 Asian Development Bank (ADB), "Annual Report 2014 – Members, Capital Stock and Voting Power," www.adb.org/sites/default/files/institutional-document/158032/oi-appendix1.pdf (accessed 23.11.2015).
71 Ibid.
72 Oliver Hensengerth, "Vietnam's Security Objectives in Mekong Basin Governance," *Journal of Vietnamese Studies* 3, no. 2 (2008): 103.
73 GMS Secretariat, "Greater Mekong Subregion."
74 GMS Environment Operations Centre, "Fast Facts – the Greater Mekong Subregion Core Environment Program and Biodiversity Conservation Corridor Initiative (CEP–BCI)," GMS, Bangkok, 2010.
75 ADB, "Greater Mekong Subregion Ministers to Move to New Phase of Environmental Cooperation," www.adb.org/news/greater-mekong-subregion-ministers-move-new-phase-environmental-cooperation (accessed 26.07.2011).
76 GMS Environment Operations Centre, "Fast Facts – the Greater Mekong Subregion Core Environment Program and Biodiversity Conservation Corridor Initiative (CEP–BCI)."
77 Ibid.
78 Hensengerth, "Vietnam's Security Objectives in Mekong Basin Governance."
79 Philip Hirsch, Kurt Morck Jensen, with Ben Boer, Naomi Carrard, Stephen FitzGerald, and Rosemary Lyster, "National Interests and Transboundary Water Governance in the Mekong," University of Sydney, Sydney, 2006.
80 Also known as the Golden Quadrangle.
81 Than and Abonyi, "The Greater Mekong Subregion: Co-Operation in Infrastructure and Finance," 155–6.
82 Ibid.
83 Oxfam Australia, "Xayaburi Dam," www.oxfam.org.au/what-we-do/infrastructure-people-and-environment/save-the-mekong/damming-the-mekong/xayaburi-dam/ (accessed 10.09.2014).

84 International Rivers Network, "The Xayaburi Dam – a Looming Threat to Mekong River," International Rivers Network, Berkeley, CA, 2011.
85 Kongrut, "China Should Set Good Example on the Mekong River."
86 Interview with a high-level expert of the MRC in Vientiane, Laos on the 06.06.2013.
87 EGAT is a Thai state enterprise that owns and manages the majority of Thailand's electricity generation capacity, as well as the nation's transmission network.
88 IRN, "The Xayaburi Dam – a Looming Threat to Mekong River."
89 Gold panning is still a big source of income in Laos, especially for women living around Luang Prabang.
90 IRN, "The Xayaburi Dam – a Looming Threat to Mekong River."
91 MRC, "Agreement on the Cooperation for the Sustainable Development of the Mekong River Basin," Article 5.
92 International Center for Environmental Management, "Environmental Assessment of Hydropower on the Mekong Mainstream – Summary of the Final Report."
93 Interview with an expert in water management in the Mekong Basin based in Vientiane, Laos on the 24.05.2013.
94 Interview with an expert in water management in the Mekong Basin based in Vientiane, Laos on the 24.05.2013.
95 Interview with an expert in water management in the Mekong Basin based in Vientiane, Laos on the 24.05.2013.
96 Douglas Varchol, *Movie: The Mekong*. CGIAR, IUCN's Mekong Water Dialogue and the Swedish International Development Agency, 2012.
97 Interview with a hydropower expert of the MRC in Vientiane, Laos on the 14.06.2013.
98 Pöyry is a Finnish-based global consulting and engineering firm focusing on energy, forestry, infrastructure and environment.
99 Interview with a water management expert in the Mekong Basin based in Vientiane, Laos on the 24.05.2013.
100 International Rivers Network, "Poeyry's Role in the Xayabouri Dam Controversy," International Rivers Network, Bangkok, 2013a.
101 Interview with a high-level expert of the MRC in Vientiane, Laos on the 06.06.2013.
102 The six-month prior consultation process for the Don Sahong Dam officially started on 25 July 2014. In early 2015 the MRC Council could not agree if the process could come to an end and transferred the decision to the respective governments for decision.
103 Guy Ziv, Eric Baranb, So Namc, Ignacio Rodríguez, and Simon A. Levina, "Trading-Off Fish Biodiversity, Food Security, and Hydropower in the Mekong River Basin," *Proceedings of the National Academy of Science of United States of America* 109, no. 15 (2012): 3; Michael A. Lerner, "Dangerous Waters: Violations of International Law and Hydropower Development Along the Sesan River," Oxfam America, Phnom Penh, 2003.
104 Interview with a hydropower expert of the MRC in Vientiane, Laos on the 14.06.2013.
105 Interview with an expert in water management in the Mekong Basin based in Vientiane, Laos on the 24.05.2013.
106 Radio Free Asia, "Laos Confirms 'Preparatory' Work on Controversial Dam Project," www.rfa.org/english/news/laos/confirms-07312013151754.html (accessed 10.08.2013); Oxfam Australia, "Lower Sesan 2 Dam," www.oxfam.org.au/explore/infrastructure-people-and-environment/save-the-mekong/3s-critical-environmental-zone/lower-sesan-2-dam/ (accessed 10.08.2013).
107 Also called the 3 S Rivers.
108 Lerner, "Dangerous Waters: Violations of International Law and Hydropower Development Along the Sesan River."

109 International Rivers Network, "Lower Sesan 2 Dam," www.internationalrivers.org/campaigns/lower-sesan-2-dam (accessed 23.05.2016).
110 *Mekong Watch*, "Lower Sesan 2 Hydropower Project, Northeastern Cambodia," www.mekongwatch.org/english/country/cambodia/LS2/index.html (accessed 01.07.2015).
111 Ibid.
112 Interview with a water management expert in the Mekong Basin based in Vientiane, Laos on the 24.05.2013.
113 *Mekong Watch*, "Lower Sesan 2 Hydropower Project, Northeastern Cambodia."
114 *Mekong Watch*, "Lower Sesan 2 Hydropower Project, Northeastern Cambodia"; Susanne Schmeier, *Governing International Watercourses – River Basin Organizations and the Sustainable Governance of Internationally Shared Rivers and Lakes* (Abingdon: Routledge, 2013a).
115 Shane Worrell, "Dam Firm Violated Law," *The Phnom Penh Post*, 05.06.2013.
116 International Rivers Network, "Lower Sesan 2 Dam."
117 International Rivers Network, "Lower Sesan 2 Dam"; interview with an expert in water management in the Mekong Basin based in Vientiane, Laos on the 24.05.2013.
118 Ibid.
119 Interview with an expert of the River Coalition Cambodia in Phnom Penh on the 06.12.2012.
120 MRC, "Agreement on the Cooperation for the Sustainable Development of the Mekong River Basin," Article 5.
121 Ziv *et al.*, "Trading-Off Fish Biodiversity, Food Security, and Hydropower in the Mekong River Basin."
122 Ziv *et al.*, "Trading-Off Fish Biodiversity, Food Security, and Hydropower in the Mekong River Basin," 1.
123 MRC, "Agreement on the Cooperation for the Sustainable Development of the Mekong River Basin," Articles 7–8.
124 Interview with a hydropower expert of the MRC in Vientiane, Laos on the 14.06.2013.
125 *Mekong Watch*, "Lower Sesan 2 Hydropower Project, Northeastern Cambodia."
126 Interview with a hydropower expert of the MRC in Vientiane, Laos on the 14.06.2013.
127 Lancang refers to the Mekong River in China.
128 Southeast Asia River Network, "Downstream Impacts of Hydropower and Development of an International River: A Case Study of Lancang–Mekong," Southeast Asia River Network, Chiang Mai, 2004.
129 Japan Environmental Council, "The Mekong Region: Incorporating the Views of Regional Civil Society," in *The State of the Environment in Asia*, ed. Japan Environmental Council (Tokyo: Springer, 2005), 134.
130 Ibid., 133.
131 Kate Lazarus, Pierre Dubeau, Channa Bambaradeniya, Richard Friend, and Latsamay Sylavong, *An Uncertain Future: Biodiversity and Livelihoods Along the Mekong River in Northern Lao PDR* (Bangkok and Gland: The World Conservation Union (IUCN), 2006).
132 Lisa Mastny, "Messing with the Mekong," *World Watch*, November/December 2003.
133 Japan Environmental Council, "The Mekong Region: Incorporating the Views of Regional Civil Society."
134 International Rivers Network, "Navigation Project Threatens Livelihoods, Ecosystem," International Rivers Network, Berkerley, CA, 2002.
135 Mastny, "Messing with the Mekong."
136 International Rivers Network, "Navigation Project Threatens Livelihoods, Ecosystem."

137 Lazarus *et al.*, "An Uncertain Future: Biodiversity and Livelihoods Along the Mekong River in Northern Lao PDR."
138 Southeast Asia River Network, "Downstream Impacts of Hydropower and Development of an International River: A Case Study of Lancang–Mekong."
139 Interview with a high-level expert of the MRC in Vientiane, Laos on the 06.06.2013.
140 International Union for Conservation of Nature (IUCN), "Upper Mekong Development," http://archive-org.com/org/w/waterandnature.org/2013-11-20_3196779_28/Huong_Vietnam_E_flows_pilot_WANI/ (accessed 29.08.2013).
141 Mastny, "Messing with the Mekong."
142 Ibid.
143 International Rivers Network, "Navigation Project Threatens Livelihoods, Ecosystem."
144 MRC, "Agreement on the Cooperation for the Sustainable Development of the Mekong River Basin," Articles 1 and 9.
145 Interview with a high-level expert of the MRC in Vientiane, Laos on the 06.06.2013.
146 *Xinhua News Agency*, "Myanmar to Host Four-Country Commercial Navigation Meeting," *Xinhua News Agency*, 05.04.2007; interview with a high-level expert of the MRC in Vientiane, Laos on the 06.06.2013.
147 Interview with an expert in water management in the Mekong Basin based in Vientiane, Laos on the 24.05.2013.
148 Interview with a high-level expert of the MRC in Vientiane, Laos on the 06.06.2013; Southeast Asia River Network, "Downstream Impacts of Hydropower and Development of an International River: A Case Study of Lancang–Mekong."
149 Mastny, "Messing with the Mekong."
150 Chris Cocklin and Monique Hain, "Evaluation of the EIA for the Proposed Upper Mekong Navigation Improvement Project – Report Prepared for the MRC – Environment Program," Monash Environment Institute, Melbourne, 2001.
151 Tyson R. Roberts, "Downstream Ecological Implications of China's Lancang Hydropower and Mekong Navigation Project," in *International Symposium on Biodiversity Management and Sustainable Development in the Lancang-Mekong River Basin* (Xishuangbanna, China, December, 2001), 9.
152 Interview with a high-level expert of the MRC in Vientiane, Laos on the 06.06.2013.
153 Interview with a hydropower expert of the MRC in Vientiane, Laos on the 14.06.2013.

# 4 Danube River: The Leading Example?

After the conclusion of the first case study, this chapter analyses the Danube River as a second case study. The Danube River is an important river traversing Central and Eastern Europe and is closely linked to European expansion and integration in recent decades. This chapter starts with highlighting the political, economic as well as ecological importance of the Danube region and the stakes of the riparian states. Power relations in the river basin are closely linked to the industrial development of the riparian countries and historical context, as can be seen between Germany and Austria on the upstream of the Danube in comparison to the former Yugoslavia and Romania on the downstream.

The second part examines the role of the European Union (EU), especially its integration and enlargement in 2004 and 2007, and the role of the aspiring riparian states with ambitions for EU membership in the near future. The research emphasises the role of EU environmental legislation, including the important Water Framework Directive (WFD), which I argue has a strong influence on the Danube River and the riparian countries. This section also focuses on the different important current multilateral agreements and cooperation in the Danube region and highlights the main institutions and their specific goals, aims and members. I focus on the International Commission for the Protection of the Danube River (ICPDR) and its influence in the region as well as its handling of environmental challenges over the last decade. I also pay attention to the traditional areas of cooperation on the Danube, mainly in the areas of navigation and hydropower development.

The third part uses cases studies within the basin to provide an insight into three key decisions made in recent times and the role that the ICPDR played in them. These include the case of the Gabcikovo–Nagymaros River Barrage System (GNRBS) along the Slovakian and Hungarian border, which was a key bilateral issue and challenge on the Danube River affecting other riparian states starting in 1977. The second case is the Baia Mare cyanide spill from a gold mine in Romania in 2000. The final case study is the ongoing issue of the Bystroe navigational canal in Ukraine.

Overall, this chapter argues that while the Danube case appears very different to the Mekong and La Plata cases, the challenges are very similar, as the three micro-level case studies demonstrate. The geographical positioning of the riparians has a strong impact on cooperation and the institutions along the Danube, and without a strong commitment by both upstream countries, the situation of the ICPDR would be very different. Additionally, the chapter shows that the ICPDR plays an active role focusing on mediation (as opposed to achieving binding legal decisions) and that the EU framework plays a major role in the cases, especially the Bystroe Canal project, and its role will only increase in the future. However, the cases also show that the implementation and control of these guidelines can be challenging and time intensive. Furthermore, although cooperation on the Danube River is well developed and public participation exists, it should be increased further, especially in the Eastern European countries as there still are enormous economic, social, environmental and political gaps between the riparian countries. Overall, European enlargement and integration plays a crucial part in the developments along the Danube River and is a key component for further successful cooperation.

## Importance of the Danube River

The Danube is Europe's second largest river, after the Volga River, and encompasses Central and Eastern European countries that share the river basin (see Map 4.1).[1] The river flows 2,780 kilometres from its source in the Black Forest of Germany to the Black Sea, touching 19 countries: Albania, Austria, Bosnia and Herzegovina, Bulgaria, Croatia, Czech Republic, Germany, Hungary, Italy, Macedonia, Moldova, Montenegro, Poland, Romania, Serbia, Slovak

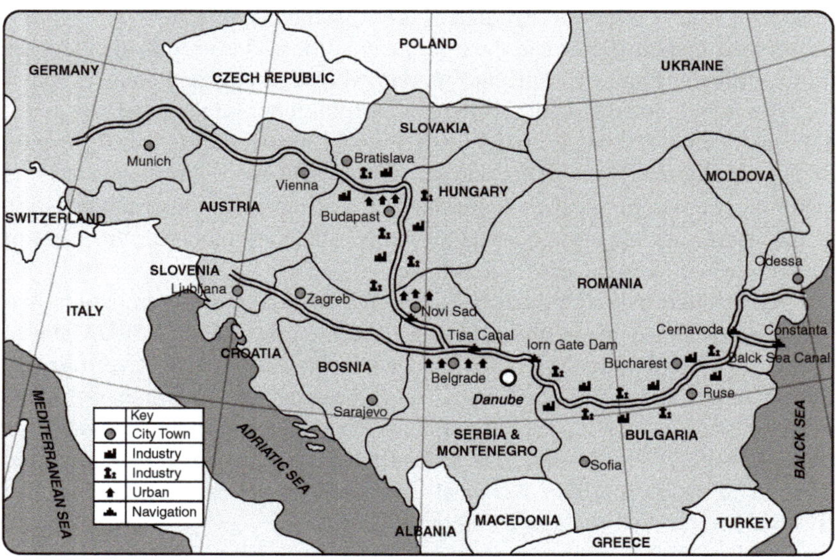


*Map 4.1* The Danube River and catchment area.

Source: Adjusted from *BBC Online*, 2012.

Republic, Slovenia, Switzerland and Ukraine. The major shared areas (proportions) of the Danube River are as follows: Romania (28.43%), Hungary (11.39%), Serbia (11.19%), Austria (9.88%), Germany (7.30%), Slovakia (5.94%) and Slovenia (2.19%).[2]

More than 82 million people of different cultures and languages are located along the Danube and are interconnected through the water system; this counts for around 10 per cent of the people on the European continent.[3] The riparian countries vary significantly in socio-economic terms, which can be seen in the Gross Domestic Product (GDP) data ranging between 2,842 and 39,454 (per capita Purchasing Power Parity (PPP) United States Dollar (USD)) in 2009, with the GDP of the wealthiest riparian country, Austria, being nearly 14 times higher than that of the poorest riparian, Moldova.[4]

The Danube has a long history of navigation, is a major shipping route and the only European river flowing from west to east.[5] The river is considered the backbone of water transport in Europe connecting Western Europe and the Rotterdam Harbour in the Netherlands with the Black Sea through the Danube–Rhine–Main Canal.[6] Overall, the Danube is navigable on 2,411 kilometres with nearly 50 per cent being border sections between two countries. The river is important to industries and factories but also vital for the large-scale agricultural production along its banks, especially in the downstream countries. Furthermore, the Danube is used for drinking water, fishing, power generation and the disposal of wastewater and sewage, which can be contentious and lead to disputes between different water users.

The Danube also has a large number of dams including hydroelectric plants along its shores with the major dams in the Upper Basin countries, Germany and Austria. So far, a total of 59 dams have been built along the river's first 1,000 kilometres – from the source of the Danube in the Black Forest in Germany down to Gabcikovo in Slovakia.[7] Developments on the Danube have influenced the water quality and quantity, which has reduced biodiversity and threatened to influence the diverse and rich ecosystem. In recent years, intensive urban, agricultural and industrial developments along the Danube River have damaged the region's biodiversity and a healthy ecosystem, with a major problem recently being non-point source pollution.

Overall, the Danube is home to important flora and fauna species; it is inhabited by large numbers of fish species, including sturgeon species, and also rare birds such as the black stork, the white-tailed eagle or the white pelican, mainly located in the Middle and Lower Basin. However, due to increasing human pressure on large parts of the river, most sections of the Danube Delta have lost their diversity and more than 80 per cent of the wetlands have disappeared.[8]

## What Is at Stake for the Riparians?

The Danube is the second largest river in Europe and has a catchment area of 817,000 square kilometres.[9] Politically, the Danube Basin went through major changes between 1945 and 1990. This is mainly due to the breakup of Yugoslavia, which occurred as the result of an ideological chasm and an increase in nationalism, which first originated in the 1970s. The country was divided into five independent nations: Slovenia, Croatia, Bosnia–Herzegovina, Serbia and Montenegro. Further, the split of the former Soviet Union into Moldova and Ukraine (amongst others) as well as the division of Czechoslovakia into Slovakia and the Czech Republic added to the countries in the Danube Basin.

These major political changes in the 1980s and 1990s influenced the riparian countries, their cooperation on the Danube Basin and affected the regional dynamics.[10] During the Cold War, the Soviet Union and the Eastern European bloc aligned against the United States (US), which was supported by its Western Allies. The Cold War influenced the Danube Basin as it divided, similarly to the Mekong countries, the riparians along ideological sides. The collapse of the former Soviet Union in 1991 resulted in the creation of newly independent countries Ukraine and Moldova as riparian countries. These political changes brought challenges and opportunities for the Danube. On the one hand, this allowed the newly created countries to participate in the free market system. On the other hand, however, this transition including industrial development drastically increased the pollution and negative effects on the Danube River resulting in lower water quality and water quantities.[11]

The Danube River Basin is divided into three subregions, namely the Upper Basin, the Middle Basin and the Lower Basin. The Upper Basin ranges from the Black Forest in Germany to Bratislava in Slovakia, including Germany, Austria and Slovakia. The Danube region has rich woodland in the Upper Basin with a forest cover of more than one third of its area; however, the Hungarian and Moldavian lowlands are rarely wooded due to deforestation as a result of producing farm and grazing land.[12] The Middle Basin is the largest of the three subregions and follows from Bratislava to the dams of the Iron Gate Gorge on the border between Serbia and Romania. The Lower Basin flows from the Romanian and Bulgarian mountains into the Danube Delta which covers an area of about 6,750 square kilometres and ends in the Black Sea. The Danube is also the main water source for Europe's largest aquifer, which is in some places deeper than 200 metres, contains ten billion litres of fresh water and covers several smaller and larger lakes, including the largest Central European lake, Lake Balaton, located in Hungary in the Middle Basin, measuring 589 square kilometres.[13]

Today, the Danube River covers 19 countries, is regulated along more than 80 per cent of its length and has been modified to help prevent floods, to obtain new land for urban development and, in many cases, hydrological engineering structures have changed the river's course. The Danube River includes 26 tributaries, the most important of which include the Inn, Tisza, Sava, Siret and Prut rivers. Overall, the Danube is of great importance to its riparian states. However, interests in water use vary enormously across the whole basin.

### *Upstream Countries*

The upper part of the Danube has been ideal for building hydropower dams due to the river's considerable natural gradient. So far, a total of 59 dams have been built along the river's first 1,000 kilometres. This means that the Upper Danube is interrupted every 16 kilometres on average and very few stretches can still be characterised as free-flowing. Germany is the source of the Danube and, with Austria, the most upstream country. Seventeen per cent of Germany, including mainly Bavaria and Baden-Wuerttemberg, is drained by the Danube River and around 9.4 million people live in the area. The Danube flows from its source in Germany around 584 kilometres until the Austrian border and loses around half of its discharge through underground passages to the Rhine River. Hydropower generation counts for around 15 per cent of total energy consumption in Bavaria and navigation plays a major role, with the Danube linked to the Rhine River by the Main–Danube Canal.[14] The Danube is important for the region it flows through; however, in a German-wide context, the dependency of Germany on the Danube River is minor as the country has several other large rivers. Nevertheless, Germany shows a high interest in protecting the river and supporting a functioning institutional framework.

The second upstream country, Austria, is home to the ICPDR and around 7.7 million people live within the Danube Basin which drains 96 per cent of the country.[15] The gradient of the Austrian section of the Danube is relatively steep and the extracted energy from running water is an important source for Austria. Around 60 per cent of the yearly electricity generation in Austria originates from hydropower in the Danube Basin.[16] The other 40 per cent is used for agriculture, settlement and infrastructure and, additionally, 350 kilometres of the Danube River in Austria is an international waterway. This shows that the dependency of Austria on the Danube is higher than Germany's.

Both upstream countries do not benefit as much as downstream countries from sustainable development on the Danube. However, Germany and Austria are drivers behind the ICPDR. This means that they are highly influential in shaping the agenda and they also contribute financially towards the ICPDR as well as to other programs supporting the less developed Danube riparians. Furthermore, besides the strong commitment, both countries are 'role models' for the other riparian states.[17]

### *Midstream Countries*

The midstream countries of the Danube are generally less developed than the upstream countries and depend more on the Danube river flow. The main usage of the Danube in the middle section is for agriculture, water supply and waste disposal. The Czech Republic sits 'on top' of Europe and around 2.8 million people live in the area of the Danube Basin within the country. Hydropower potential is low and only 4 per cent of energy production is generated by hydropower plants.[18] Additionally, the waterways do not meet the requirements for international navigation on the Danube River and only small navigational activities for recreational purposes exist.[19]

The most significant activities that may impact on the morphological alternation are the significant flow regulation, hydropower generation, water abstraction for urban, industrial and agricultural uses and flood defence measures. Additionally, past uranium mining activities are a potential pressure regarding groundwater quality in the Morava River Basin, a sub-basin of the Danube.[20] Ninety-six per cent of Slovakia is within the Danube Basin and around 30 per cent of energy production derives from hydropower, including energy from the Gabcikovo dam.[21] Regarding navigation, the Danube River is the only important international waterway in Slovakia.

Hungary is located in the middle of the river basin and is highly concerned with pollution on the river. Hungary's entire territory contributes to the basin and also inhabits the largest lake of the Danube Basin, Lake Balaton, comprising two million cubic metres with a catchment area of 5,775 square kilometres, which is a key recreational area for Central and Eastern Europe.[22] Hungary is very dependent on upstream countries for water with a dependence ratio of around 96 per cent[23], a situation comparable to Cambodia and Paraguay in the other two case studies. Additionally, flooding is a major issue in Hungary, especially fast-rising ice-jam floods. Hungary does not have significant hydropower potential and only around 1 per cent of energy production is generated by hydropower. Large parts of Hungary (around 1,600 kilometres of waterways) are navigable, but cargo transportation is very small. Another problem is that the two major rivers are not connected to each other, an obstacle to the integration of inland waterway shipping into the national transport economy. Overall, the Danube and Tisza (a sub-basin of the Danube) rivers play a central role in Hungary and are often seen as intertwined with Hungary's existence as a nation-state, reflecting a historical preoccupation with borders and territorial integrity.

Slovenia's interest is a sustainable Black Sea, which is linked to the Danube River. Additionally, the Sava River is the first tributary of the Danube with a sub-basin organisation (Sava River Basin Organisation) and Slovenia is a key player in it, placing an emphasis on sub-basin activities.[24] Croatia is a Central European Mediterranean country and around 62 per cent of its mainland is located in the Danube Basin.[25] Croatia's hydropower potential is mostly exhausted, besides a few valley locations which would, however, potentially cause great environmental impacts affecting groundwater and surface regimes. Major international waterways include the Danube and the Drava Rivers (a tributary of the Danube) and more than 80 per cent of goods shipped on the inland waterway in Croatia involve international transport, which demonstrates its importance. Similar to Hungary, Croatia also has a high flood risk and has been affected by large floods in recent years.

The Sava Basin is the largest sub-basin of the Danube River and lies to around 40.2 per cent within Bosnia and Herzegovina; large parts of the watercourse are international waterways. Thus far, 12 hydropower plants are located on the tributaries, mainly on the borders with other countries.[26] Serbia is located to around 92 per cent within the Danube Basin and approximately 90 per cent of available water originates outside the country, making it very water dependent and turning international water cooperation into a vital issue. The main usage of the Danube water in Serbia is for domestic and industrial water supply, irrigation, navigation and the cooling of thermal power plants. Hydropower is a significant power generator in Serbia, accounting for 31 per cent of total production in 2004 including the biggest dam projects on the Danube, the Iron Gate I and II dam systems.[27] Navigation exists on the 1,700-kilometres-long inland waterway network connecting the Danube, Sava and Tisza rivers. This network is linked to the European inland network, the extension of which is one of the key purposes for Serbia in the future.

### *Downstream Countries*

The downstream countries are the least developed riparians along the Danube River and use the river for different purposes than the upstream and midstream countries. Whereas Romania and Bulgaria joined the EU in 2007, Montenegro is a candidate country and Moldova and Ukraine have currently no status in the EU.

Montenegro's biggest natural resource is water, and around 95 per cent of the waterway in Montenegro originates from within.[28] As a result, hydropower is Montenegro's most important energy resource potential and currently only 17 per cent of the potential is used, which shows the future potential in the area in a similar fashion to Laos on the Mekong. Romania is nearly entirely situated within the Danube Basin and is the end carrier of all wastewater discharge from upstream countries until it flows into the Black Sea. The country is poor in water resources with 1,700 cubic metres per inhabitant per year, ranking 13th in a European comparison.[29] Hydropower accounts for 30 per cent of national generation capacity in Romania, which has more than 400 large dams. Furthermore, chemical fertilisers used in upstream countries and the lack of proper functioning wastewater treatment plants in Central and Eastern Europe affect the water quality within the Danube in Romania before ending in the Black Sea. Romania's desire to obtain EU membership represented an opportunity for the EU to provide stability to the lower Danube region, which was achieved in 2007, moving the geopolitical interest of the EU to the Ukraine. The Danube River catchment basin comprises around 42.4 per cent of the territory of Bulgaria and totals 46,930 square kilometres, encompassing almost all of northern Bulgaria.[30]

Moldova hosts around one million people along the Danube River Basin and 80 per cent of the area is dominated by agricultural use. For flood prevention, around 90 per cent of the river bank is dammed at an average height of two metres. Energy production is very low and only one hydropower station exists in the Danube River Basin in Moldova.[31] The Danube flows through the lower parts of the Ukraine and encompasses one of the largest wetland areas in Europe, including more than 5,000 animal species such as dolphins, seal-monks, pink pelicans and geese. The Ukraine had access to Europe through navigation on the Danube in the past, but today the waterway is only navigable for small ships. Navigation is a central part of the efforts in the Ukraine along the Danube to improve navigation and reduce the sharp decline in cargo shipped through the Ukrainian ports. The disputed key project is the Bystroe Canal with a planned length of 162.6 kilometres and an estimated cost of 30 million euros (EUR).[32]

### *Conclusion: Riparian States*

Due to the upstream–downstream phenomena, there exist few possibilities for trade-offs for downstream countries to achieve upstream favours in the Danube.[33] Some argue that if the Danube flowed in the other direction, cooperation would be easier to achieve as trade-offs and side payments would be possible when the dominant countries are located downstream. However, although the geographical positioning is uneven, with the most developed countries located upstream and the least developed downstream, due to the strong commitment and the 'role model' behaviour of upstream countries Germany and Austria and other regional incentives (mainly EU accession), cooperation is working well on the Danube River Basin.

## Regional Developments – The Role of EU Integration for Cooperation

Nearly all riparian countries involved in the Danube Basin and member countries of the ICPDR are in the EU, in the process of joining or are influenced by developments within the EU. As a result, it is important to understand the background and developments in the EU and how these influence the development on cooperation and progress in the Danube Basin and vice versa. A main focus is on the EU Enlargement, mainly the fifth and sixth enlargement phase in 2004 and 2007, which included the riparian countries Czech Republic, Hungary, Slovakia, Slovenia, Romania and Bulgaria. Further, this section will concentrate on the riparian countries Croatia (the most recent EU member who joined in 2013), the 'candidate countries' Montenegro and Serbia and the 'potential country' Bosnia–Herzegovina. Lastly, this section will highlight some of the important environmental changes due to EU legislation, with a strong focus on the WFD and its influence on new and existing EU member countries.

### *The European Union (EU) – A Short Timeline and Enlargement*

The EU was founded in 1951 comprising six countries – Belgium, Germany, France, Italy, Luxembourg and the Netherlands as the European Coal and Steel Community (ECSC) – with the goals of establishing long-lasting peace in

Europe and fostering economic growth.[34] While continuously growing, the biggest EU Enlargement took place in 2004 when the EU expanded from 15 to 25 countries. The main reasons for the Central and Eastern enlargement included the stabilisation of the region's new political and economic systems; to assist Europeans competing in a globalised economy; and to improve the environmental protection and quality in the new joining countries but also in continental Europe as a whole. This historic enlargement also included four of the Danube riparian countries, namely the Czech Republic, Hungary, Slovenia and Slovakia.[35] The next enlargement was in 2007 and included the riparian countries Romania and Bulgaria, whilst Croatia gained access in 2013.[36] So far, the enlargement rounds in 2004 and 2007 have had the biggest implications for the Danube River in regards to cooperation, implementation of environmental standards and guidelines, and financial support.

Joining the EU is a top priority for the Balkan countries such as Albania, Bosnia–Herzegovina, Macedonia, Montenegro and Serbia. Similarly, most of the former Yugoslavia countries are still fragile, except Slovenia, which joined the EU in 2004 and Croatia in 2013. All the other countries are still moving towards European integration; however, corruption is still a problem in government and public services.[37] Further, environmental legislation and environmental awareness have been very low in these newly independent countries such as Montenegro. However, if these countries want to join the EU, they must comply and enforce strict environmental legislation including challenges such as reducing pollution from hazardous waste rock, tailing dams, mine voids, open pits and other hazards, which will be highlighted in the micro-case analysis.[38]

### *EU Environmental Legislation – A Brief Overview*

During the creation of the EU and in its first decades, environmental matters have not been dominant and were enacted mainly due to concerns about trade and competition and the impact of pollution on human health. However, this changed in the 1970s and 1980s due to a better scientific understanding of environmental problems and an increase in public interest. Also, headlines of environmental disasters, including concerns about acid rain in the early 1980s, the Chernobyl nuclear accident in 1986 and the toxic waste spill polluting the Rhine River in the same year increased publicity.[39]

A first step was the Environmental Action Program of 1973, which was followed by key environmental laws in the EU including the Single European Act signed in 1986, entering into force in 1987 when the competency of the European Community (EC) in environmental policymaking increased and environmental protection was added to the responsibilities of the EC, including the Polluter Pays Principle. More specific laws include the 1988 law of the reduction of acid rain incidence, and the cutting of emissions of sulphur dioxide and nitrogen oxides by 58 per cent over the coming 15 years.[40] Two other important laws in regards to river basins include the introduction of standards for the protection against hazardous chemicals in 1995 and the introduction of sustainable development in the 1990s, which included the regulation that the impact on the environment needs to be taken into account before an economic development should be agreed on. Also, the acceptance of the Polluter Pays Principle as community law in 2004 played a key role, especially in regards to river basins, as the polluter often was not the one affected by the damage and was not responsible for compensation.

As highlighted in the Copenhagen Criteria, one of the key prerequisites for EU membership is the obligation to apply all EU policy and legislation. In the environmental sector this is a shift towards a cleaner environment. However, this also implies a major burden for potential EU member states as there are, for instance, over 300 pieces of EU environmental legislation that have to be transformed into national legislation (including air and water pollution, noise limits for motor and aircraft vehicles, waste disposal, wildlife habitats and quality standards for drinking and bathing water).[41]

Often weak and under-resourced administrations need to be strengthened in order to comply with the environmental standards. Further, substantial investment in technology and infrastructure is needed to achieve minimum levels.[42] Overall, the Environmental Chapter of the EU – which every new member state needs to fulfil – is one of the most difficult ones.[43] This is also accompanied by the European Commission's Opinion on the individual countries such as of the candidate countries Serbia and Montenegro that have similar problems in regards to the Environmental Chapter. For example, Montenegro is currently at a negotiation stage and screening process where the EU and the country assess the current development status and determine what further improvements are necessary. One of the problems is that Montenegro is relatively small and, therefore, the Environment Chapter is a big burden with a lot of work, including the alignment of the national legislation with the EU legislation and the implementation of these changes. The cost for adopting the environmental acquis is estimated to range from 80 to 100/110 billion EUR, which is, on average, 2–3 per cent of the GDP of the candidate countries. This needs to be financed primarily from the countries' own resources.[44]

Overall, however, the potential future EU countries are at different stages, which means that there is leverage in regards to the environmental criteria. Furthermore, the EU also pushes for closer cooperation between the aspiring countries so that they can learn from each other during the different stages of the EU membership process.

### *The EU Water Framework Directive (EU WFD)*

One of the major challenges, not only for new EU member states but also for the traditional EU countries, is the WFD which deals with water management on river basins directly impacting on the Danube. Early European water legislation – covering rivers and lakes – included the drinking water abstraction in 1975 followed by binding quality targets five years later. Already in 1988, the Community had stated that additional improvements of ecological quality in surface waters were necessary, something that was developed further in The

Hague in 1991. There, it was recognised that action is needed to avoid the long-term deterioration of freshwater, in regard to both quality and quantity.[45]

A strong focus was placed on the precautionary principle, that preventive action should be taken and the importance of the Polluter Pays Principle was highlighted. In 1991, the Community adopted the Urban Waste Water Treatment and the Nitrates Directive, which focused on the water pollution from the agricultural sector, followed by the new Drinking Water Directive and the Directive for Integrated Pollution and Prevention Control adopted a few years later, all having a direct impact on national and international river basins. These ideas and goals were further developed into the WFD, which came into force in 2000 covering the protection of inland surface waters, transitional waters, coastal waters and groundwater. The WFD focuses on sustainable water use based on a long-term protection of available water resources, aims at enhanced protection and improvement of the aquatic environment, ensures the progressive reduction of pollution of groundwater and prevents its further pollution.[46]

The WFD also includes several definitions with measurements and requirements, which need to be achieved by the member states. This is especially difficult for the new EU member states and the aspiring states, as it requires advanced technical equipment, educated staff and financial resources. Further, the WFD also established the River Basin Management approach, which means that the water systems need to be seen as a whole geographical and hydrological unit and that every river basin in the EU needs to provide a River Basin Management Plan, which needs to be updated every six years; this was done in the Danube Basin in 2009 with an updated version in 2015.[47] Another important step of the WFD, which had a significant influence on the riparian states, was the role of public participation, which means that during the WFD process all interested parties such as water users, Non-Governmental Organisations (NGOs), local and regional authorities, industries and agriculture bodies were involved. Public participation was also supported in the River Basin Management Plan: in the Danube case, several observers commented on progress and initiatives, something often seen as one of the key drivers of successful cross-sector cooperation.

Overall, this section shows that EU development and enlargement are directly linked to the riparian countries in the Danube River and that the EU legislation has a direct impact on the basin. Until today, the influence of the EU in regards to environmental legislation and awareness places a big onus on the existing and possible future EU countries and provides new challenges and opportunities for cooperation on the Danube.

## Cooperation on the Danube River

Treaties have influenced the Danube and the use of the river for almost 200 years. Cooperation through institutions started on the Danube River as early as 1856 with the European Danube Commission (DC) in the field of navigation, which was based on the Treaty of Paris. However, non-navigational uses were not addressed until the end of World War I through the Treaty of Trianon in 1920, which covered resolution disputes in areas such as fisheries, waterpower, irrigation and other national interests.[48] Two decades later, in 1948, the Belgrade Convention which created the DC was signed with its 11 members focusing on free navigation on the Danube. The Belgrade Convention gave control regarding navigation to each riparian; semi-legislative powers for navigation and inspections were also given to the European Commission of the Danube.

Due to geo-political changes in the region in the 1980s–1990s, resulting in a democratic and economic transition of the Eastern European countries, the Danube became more and more polluted due to increased industrialisation of the countries and the by then eight riparian countries signed the 'Declaration of the Danube Countries to Cooperate on Questions Concerning the Management of the Danube' in 1985. Also called the Bucharest Declaration, this was the first time the problem of water quality and management challenges in the Danube Basin were addressed collectively.[49] In the following years pollution was still a major problem on the Danube and an increased involvement of the EU followed, which provided financial and institutional support to the riparian countries in different projects and programs combined with the support of the UN Development Program (UNDP), the World Bank and the later-established European Bank for Reconstruction and Development.

This was followed by an intergovernmental coordinating body, the Environmental Program for the Danube River Basin (EPDRB) in 1992. These programs and processes resulted in the Program Coordination Unit which developed the Strategic Action Plan in 1993 focusing on public participation in all countries on all levels, building the base for the 'Convention on Cooperation for the Protection and Sustainable Use of the Danube River' (the Danube River Protection Convention). The key focus of the Convention was the integrated water management process and the continuous support of public participation resulting in the transnational body, the ICDPR.

## International Commission for the Protection of the Danube River (ICPDR)

The ICPDR is based on the Danube River Protection Convention,[50] which is the major legal instrument for cooperation on transboundary water management in the Danube River Basin. It was signed in 1994 and ratified in 1998. All countries sharing over 2,000 square kilometres of the Danube River Basin and the EU are contracting parties of the 1994-established ICPDR.[51] This includes 14 riparians – Austria, Bosnia and Herzegovina, Bulgaria, Croatia, Czech Republic, Germany, Hungary, Moldova, Montenegro, Romania, Slovakia, Slovenia, Serbia, Ukraine – and the EU.[52] Out of the 14 riparian countries of the ICPDR, nine are EU member states. Further, Montenegro and Serbia are 'candidate countries' and Bosnia–Herzegovina is at the stage of 'potential candidate'.

Moldova and Ukraine do not have any official application with the EU and, therefore, are not directly linked to EU legislation and policy; however, they have close ties with their neighbour countries and also can receive funds from the EU in order to move closer to EU legislation standards.

Additional countries on the Danube – not members of the ICPDR – include EU members Italy and Poland and non-EU members Switzerland, Albania and Montenegro. All five countries have only minor shares on the Danube and no interest in the water governance of it. This is no major hindrance for the ICPDR itself or cooperation along the Danube; this is contrary to the Mekong River where the absence of the two upstream countries is a major challenge. These nations still have ties with the ICPDR countries through the EU WFD, with the EU members Italy and Poland and through bilateral ties with the other non-EU countries.

Overall, the Danube Convention supports different actions on the river to improve the environmental situation of the Danube with the overarching goals being:

> Sustainable and equitable water management, including conversation, improvement and the rational use of surface waters and ground water in the catchment area as far as possible. [. . .] [Parties] shall cooperate on fundamental water management issues and take all appropriate legal, administrative and technical measures, to at least maintain and improve the current environmental and water quality conditions of the Danube and of the waters in its catchment area.[53]

Furthermore, the Danube Convention includes the Polluter Pays Principle and the Precautionary Principle. Additionally, as with all European waters, the Danube River is managed by using a river basin approach since the EU WFD was adopted by the EU in 2000. The ICPDR is the platform for the coordination of issues of international importance for the implementation of the EU WFD and supports the cooperation between the Danube River Basin countries towards the implementation of the EU WFD.[54]

### *The ICPDR – Secretary, President and Funding*

The ICPDR is divided into the Permanent Secretary and the President. The Secretariat was officially opened in 1999 and is located in the premises of the United Nations (UN) in Vienna, Austria. The Executive Secretary leads the Permanent Secretariat and is responsible for the general management and supervisory functions of the staff and related tasks (see Figure 4.1).[55] The main duties of the Secretariat are to support the work of the ICPDR and its expert bodies, to coordinate the ICPDR's work, to support project development and to maintain the ICPDR Information System.[56] Technical and administrative staff supervise and control the quality of the Secretariat's main functions and tasks, including project staff.[57]

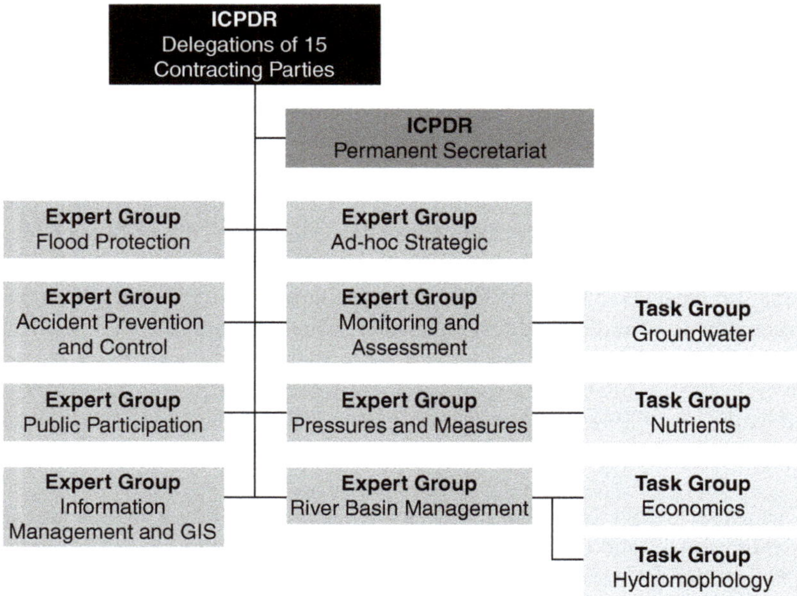


*Figure 4.1* Organisational structure of the ICPDR.
Source: adjusted from ICPDR, 2014.

The key task of the ICPDR in regards to water management on the Danube is to coordinate the riparian countries and provide a discussion forum for them. This also includes the involvement of other active non-state actors with observer status. There are over 20 observers of the ICPDR which include NGOs, organisations representing private industry and intergovernmental organisations. Some of the more influential observers are: the World Wide Fund for Nature – Danube–Carpathian Program (WWF–DCP), Via Donau, the Regional Environmental Centre for Central and Eastern Europe (REC), the European Anglers Alliance (EAA), the European Barge Union (EBU), the DC and the International Association for Danube Research (IAD).

All observers are invited to the meetings and can provide inputs to the proposals and raise awareness for their projects and goals in regards to the Danube. Observers are also involved in all recent policy decisions such as the Basin Management Plan 2009 published by the ICPDR – one of the biggest milestones in the Danube Basin cooperation. Many experts argue that the observer status is one of the key reasons for the successful work of the ICPDR and raises the necessary public awareness to support the developments on the Danube. It often also achieves a balance between different ideas and opinions due to the representation of a wide diversity of actors involved in the process.

Besides the Permanent Secretary, the ICPDR is officially represented by the President. The Presidency of the Commission changes in alphabetical order among the riparian countries annually and the country of the presidency nominates a president. In 2012 Austria nominated Wolfgang Stalzer, who was the first Executive Secretary of the ICPDR for several years, as president. This provided major influence for Austria and its goals for the ICPDR due to the experience of Wolfgang Stalzer and his knowledge of the organisation and the Danube itself.[58] Overall though, the influence of the presidents varies and the input differs from a low contribution and interest to a very high importance and involvement, depending on the nomination from the riparian countries. Additionally, it is a goal of the ICPDR that the previous, current and future presidents work together to agree on a continuous work program for the Commission and to focus on a few key points during their country's leadership instead of changing the Commission's focus annually. In general, as highlighted by an internal expert, 80 per cent of the work schedule of the ICPDR is continuous, whereas one or two projects are different each year.[59] In 2012, for instance, Austria focused on hydropower development and securing the funding for the next Danube Survey.[60]

The ICPDR is funded by the riparian member countries equally, which means that all the countries provide the same amount of money, 100,000 EUR, annually to run the ICPDR secretary, independent of the GDP and size of the countries. However, due to financial difficulties in some countries, currently there are four countries suspended from paying the whole amount. These include Bosnia–Herzegovina, Moldova, Montenegro and Ukraine which pay 1 per cent of the total budget, an amount that will increase over time with the goal of equal payment among the riparian states.[61] Besides funding the Secretariat, the riparian states pay for the country's specific personal costs, meeting settings and organisation as well as hospitality, rooms and other costs involved in the participation at working groups or annual meetings. Overall, this means that for the poorer countries in the region participation and cooperation on the Danube River is a costly and often challenging burden. However, equal funding is seen by many experts as one of the keys to successful cooperation.[62]

### *ICPDR – Working Groups and Projects*

The members of the ICPDR and their representatives meet annually with one additional meeting. Besides this, there exist other bodies of the ICPDR including the Expert Groups and the Tasks Groups (see Figure 4.1). The working groups often have a regular number of people to prepare the proposals for the annual meeting. These groups are a key part of the ICPDR's work and include the Expert Group on River Basin Management, the Flood Prevention Expert Group and the Public Participant Expert Group, consisting of members of the contracting parties and observers.[63] In general, the decisions of the working groups and the annual meeting are accepted with consensus.

In the early years of the ICPDR, the focus was on the reduction of pollution and on water chemistry – the traditional approach to water management – which only considers water instead of including the land as well. In recent years this has shifted to a more inclusive basin approach, with the WFD adopting the notion that cooperation on water resource management can only be successful if a basin-wide approach, including water, land and flood restoration, is used. Today, some of the key issues of the ICPDR include pollution from non-point sources, especially from agriculture and the industry sector, and restoring the natural river structure. However, in general there are several other issues and challenges that need to be addressed, including possible future threats resulting from climate change. Overall, the commission works as a coordinator to bring the riparian countries together in order to exchange ideas, information and progress. The goal is to achieve a comprehensive basin-wide approach. Furthermore, the ICPDR tries to provide a forum for other bodies such as the navigation, hydropower, environmental protection or fishery lobbies.

Overall, this section showed that upstream–downstream relations have a strong impact on the management of water resources. They also have an impact on the performance of the ICPDR as the dominant power relations influence the agenda and focal points of its work. The ICPDR and its agenda would be different if the two upstream countries did not support it as much as they have. Recent movements among ICPDR signatories, however, show a transition, including the development of a revised agenda by the previously less dominant countries which could block the upstream countries agenda in the future and lead to possible future challenges between the riparians.[64] Additionally, there have been discussions in the last few years about who holds sway in the ICPDR, with strong personalities playing a major role. Yet, this can be seen as a positive sign and a stronger interest of the mid- and downstream countries. Overall, it is unclear whether the ICPDR will change as the previously less developed countries become 'visible' and could build a block to influence the agenda and direction of the organisation. It remains to be seen what impact this would have on the functionality and effectiveness of the ICPDR and developments on the Danube River in the future.

Besides the ICPDR, there exist several other intergovernmental organisations and groups that have played a key role in the development of the Danube. These include the DC focusing on free navigation and the International Hydrological Program (IHP) concentrating on hydrological cooperation.

## Navigation on the Danube – Danube Commission (DC)

Cooperation around navigation has a long history on the Danube and cooperation has expanded in recent decades into new areas including integrated water management and sustainable environmental development.[65] Navigation on the Danube Basin is based on historical experience of navigation control on the international rivers of Europe, including the European DC created under the Paris Peace Treaty of 1856. This established, for the first time, an international regime to safeguard free navigation on the Danube, and was revived in 1921.[66]

The international DC is an intergovernmental organisation which was established in Budapest under the regulations of the Belgrade Convention of 1948; it regulates the navigation on the Danube River as an international waterway. The DC was established by the Convention regarding the regime of navigation with a main focus on the free navigation on the river for commercial vessels from the riparian countries. Another of its objectives is to develop and increase economic and cultural relations between the Danube states and other countries. The current 11 member states are Austria, Bulgaria, Hungary, Germany, Moldova, Russia, Romania, Serbia, Slovakia, Ukraine and Croatia. The seat of the DC has been in Budapest since 1954 and the official languages are German, Russian and French.[67]

The goal of the DC is a unified navigation system of inland waterways in Europe, which supports safer navigation and the integration of the Danube as a significant and reliable transport corridor. The tasks include providing and unifying mutual recognition of the basic regulatory documents necessary for the river navigation on the Danube; and also, close cooperation with other international bodies and groups dealing with inland waterway transportation such as the European Commission or the Central Commission for the Navigation of the Rhine. To do so, the DC meets twice a year and also has expert groups whose main tasks include the implementation of the 1948 Belgrade Convention. This means preparing a general plan of the main work in regards to navigation interests, establishing unified traffic regulations on the whole navigable part of the Danube, customs and sanitary inspections, and regulations governing the river.[68] The president and the secretary of the DC are elected from among its members every three years. The Secretariat currently has ten staff members and is led by Director General Petar Margić (in 2015).

Navigation on the Danube has existed for hundreds of years, but has been interrupted by several incidents and disasters. For instance, navigation was closed for several years as a result of the North Atlantic Treaty Organisation (NATO) bombing of Serbia in 1999 which resulted in destroyed bridges blocking the river. Further, in recent years, 2010 and 2011, navigation on the Danube was stopped due to low water levels or ice, costing an estimated 6 million EUR and affecting the image of reliable river navigability.[69] As a result, in June 2012, the ministers from the Danube countries signed a new agreement reaffirming the importance of full navigation, which should prevent incidents that stop traffic on the Danube.[70] The ministers of Austria, Bulgaria, Croatia, Germany, Moldova, Romania and Slovakia signed the declaration, while Serbia and Bosnia–Herzegovina supported it from a distance. However, the biggest challenge is that Hungary – which has a large stretch of navigational waterways and is located in the midstream of the river – did not sign the declaration, which could hinder the success of the agreement in the future.

## International Hydropower Program (IHP)/United Nations Educational, Scientific and Cultural Organisation (UNESCO) – Hydrological Development

Besides navigational matters, hydrotechnical development on the Danube started as early as the beginning of the 19th century, including systematic river training works to prevent health problems such as epidemic fever or the protection of agricultural areas under the DC. Later, river channel training works, dikes, barrages, reservoirs and diversions were installed and influenced the natural riverbed in order to achieve navigation, flood protection, drainage, irrigation or electric power.[71] The working group for scientific hydrology was founded in 1967 and expanded the previous work to other themes such as water balance, besides the already tackled issues of evapotranspiration, river training or storage reservoir dimensioning. However, during that time, only four out of the eight countries in the DC (namely Czechoslovakia, Hungary, Bulgaria and the Soviet Union) sent experts to the scientific hydrology group.

As a result, the first working papers and measurements were unsatisfactory due to non-compliance of the other riparian countries and inconsistencies between the national data. This was followed by hydrological cooperation on the Danube River between all the eight riparian countries on voluntary bases since 1971.[72] The aim was to produce continuous hydrological data for the whole Danube Catchment. In 1974, a framework was established which included all the eight Danube countries, with German and Russian selected as working languages. Hydropolitics were still the key focus and within it the main concentration was on the publication of the 'Hydrological Monograph of the Danube Catchment', which was finally published in different languages between 1986 and 1989. During that time, it was also agreed that an institutional continuation of cooperation was necessary and this was placed under the aegis of the IHP under the United Nations Educational, Scientific and Cultural Organisation (UNESCO) in 1987.[73] This institutionalisation was supported by the IHP programs of Germany, Austria, Slovakia and Hungary and the other riparian states, resulting in the agreed upon and accepted Principles (Constitution) in 1987 and its update in 1996.

These mostly still valid principles include the goal of joint cooperation on scientific themes, but also on other common interests concerning the whole Danube Basin. Further, the constitution includes the promotion and realisation of an annual meeting of the riparian countries, and a voluntary chief coordinator that should be an expert or institution from one of the Danube countries and accepted by all members.[74] The main two focus areas are project works and conferences. Expert Working Groups meet annually to discuss the projects and attend regular conferences. The key projects of the cooperation under the IHP include the update of the Hydrological Monograph of the Danube Basin, the thermal and ice regime of the Danube River, flood coincidence and long-term fluctuations of precipitation.[75]

## Other Actors in the Danube Basin

Besides the above-mentioned transnational bodies, there exist several other active non-state actors in the Danube region – including NGOs, organisations representing private industry, and intergovernmental organisations – that support different goals and activities. Most of these groups are involved in legislation drafts and have influence on the developments in the Danube Basin. The observers can comment on proposals and texts and can also provide ideas and statements.

The World Wide Fund for Nature (WWF) is one of the more active and influential players regarding Danube development focusing on environmental impacts. The main focus of the organisation is threefold, namely the floodplain restoration on the Middle and Lower Danube, the rejection of unsustainable navigation projects, and the conservation of sturgeons.[76] However, an increasing focus is also being placed on hydropower, which the WWF sees as a new threat to the Danube Basin and its tributaries. Besides the WWF, the International Association of Danube Research (IAD) and European Anglers Alliance (EAA) also influence Danube developments.

The IAD was founded in 1956 and is an association based on Austrian law. It is the longest existing international scientific network in the Danube region. Its goal is the promotion and coordination of activities in the fields of limnology, water management, water protection and sustainable development in the Danube River Basin.[77] The EAA was launched in Brussels in 1994 and is a pan-European organisation. There are more than three million affiliated members from the 14 EAA member organisations. The EAA is recognised as an NGO and is the only European angling organisation of its kind defending recreational anglers' interests at the European level and beyond.[78] Further, the hydropower supporters have a strong lobby behind the scenes and, therefore, are also relatively influential in regards to the developments on the Danube.[79]

NGOs play a crucial part in the Danube region, although they are developed differently in the riparian states. The Environment Forum for NGOs includes several NGOs from the different aspiring countries who work together and exchange experiences. This forum is strongly supported by the EU, which emphasises NGO participation during the EU membership stages of possible EU member states. In practice this means that if NGOs cannot access their own government or public opinion, the EU receives the message and will put it forward into the country.[80] The role of NGOs is especially important in the environmental sector because they can be used as an observer and 'supervisor' within the countries. Further, during the EU membership stages, as well as under the WFD, Environmental Impact Assessments (EIAs) of large projects play a crucial role and NGO participation, besides other activist groups, is legally necessary. Many experts argue that this is one of the key reasons for the successful work of the ICPDR as it often achieves a balance between different ideas and opinions due to the representation of a wide diversity of actors being involved in the process.

Moreover, in order to understand the overall picture, it is important to include, besides the regional framework of the EU and the basin-wide institutions such as the ICPDR, the micro level and examine specific case studies to see how guidelines are implemented or not and how the three levels of analysis are interlinked.

## The Gabcikovo–Nagymaros Project

One of the largest and most controversial projects on the Danube River and in Europe remains the GNRBS, a barrage system with large dams at Gabcikovo, Czechoslovakia (later Slovakia) and Nagymaros, Hungary (see Map 4.2).[81]

This project was proposed as early as 1951, started in 1977 and was strongly supported by the Soviet Union who wanted to increase the quantity of goods shipped through this part of Eastern Europe, providing direct access to the Black Sea from Budapest and to the North Sea from Bratislava. Although the case goes back for several decades, it is a good example of potential conflict in river basins today as the key rationale for the project – including improved navigation, cleaner energy generation and flood control – remain present challenges. The case also shows how a bilateral treaty can become a problem for other riparian states, in which external moderators – such as the EC or the International Court of Justice (ICJ) – are necessary in order to find a solution. This example of the GNBRS shows that although the conflict arose more than 30 years ago, it can happen in river basins today as the challenges are still acute with potential to trigger disputes between states.

In September 1977, socialist countries Hungary and Czechoslovakia signed the 'Treaty Concerning the Construction and Operation of the Gabcikovo System of Locks' that entered into force in June 1978 with work starting in the same year. The treaty included a construction of dams, hydroelectric plants and a diversion canal on the Danube with a length of nearly 200 kilometres, including a 52-square-kilometre reservoir in close proximity.[82] Goals of both countries

*Map 4.2* The Gabcikovo–Nagymaros hydropower project (original version).

Source: Adjusted from UN Watercourses Convention, 2014.

included improved navigation, flood control and energy self-sufficiency (especially during the global oil crisis in the 1970s).

The agreement included four key areas: electricity production, navigation improvement, flood protection and regional development. According to the treaty, both countries wanted to use the jointly owned and operated barrage system to 'strengthen the fraternal relations of the two states and significantly contribute to bringing about the[ir] socialist integration'.[83] However, four years later Hungary faced economic difficulties and pressed for a temporary termination of the project. Work was slowed down on the project and, at the same time, environmental protests within Hungary against the GNBRS emerged.

At the same time, Austria and Czechoslovakia planned to build a dam (the Hainburg Dam) between both capital cities, Vienna and Bratislava. Official protest from Czechoslovakia about the disruption of the ecological balance and the danger of flooding combined with the domestic pressure from conservation groups such as the WWF and the Austrian Green Party stopped the Hainburg Dam project, resulting in no work for the Austrian construction firm and leaving the investors without a project.[84] As a result of the failed project, Austria offered its help to Hungary to work on the Nagymaros Dam – including the supply of investment and Austrian workers – in exchange for a large share of the electrical output to be transferred from Hungary to Austria for the next 20 years.

As a result of this offer from Austria, Hungary was under pressure from the treaty obligations with Czechoslovakia, which meant that there was no other option than to restart work on the dam despite domestic economic pressures in Hungary. However, non-compliance would have cost Hungary around 55 billion Hungarian forints, including 30 billion forints to Czechoslovakia, 8 billion forints to Austria for loss of investment and 17 billion forints to restore the Nagymaros site.[85] However, in 1989, after the revolution, the newly democratic elected Hungarian government withdrew from the project because of increased opposition and as a precautionary measure, as possible transboundary damage was anticipated. As a result of the withdrawal from Hungary, the change in the system in 1989 and the dissolution of Czechoslovakia, Slovakia decided to implement a new variation of the project. This variation would redirect the Danube into a new canal at a position near Gabcikovo, instead of the original position close to Nagymaros, also called Variation C (see Map 4.3).

This change would, after completion, extract around 90–97 per cent of the Danube water from the old riverbed to Gabcikovo.[86] Slovakia, located uphill of Hungary, focused on economic survival and development, and ignored the protests of Hungary in order to finish the dam project. Slovakia was more heavily industrialised than Hungary in the late 1980s and early 1990s and needed the additional water forecasted through the dam project. Further, the dam construction was seen as an attractive energy option for different reasons: it was the cleanest large-scale source of electricity and a non-polluting and renewable option, providing electricity for millions of people at a low cost with minimal impact.[87] Moreover, the project created new jobs for building and maintenance in the local economy.

*Map 4.3* The Gabcikovo–Nagymaros hydropower project (Variation C).
Source: Adjusted from UN Watercourses Convention, 2014.

Slovakia ignored the drastic effects downstream in Hungary, which included a two-metre water-level drop. In a similar fashion to the other key decisions on the Danube River and the other two river basin case studies, the Gabcikovo project construction went ahead mostly due to the support of the ruling elite and Slovakia's nationalist rhetoric focusing on industrialisation and national development. Tensions increased between the two countries over the 1977 bilateral treaty, resulting in an international dispute and the involvement of the EC and the ICJ in 1993.

However, the ICJ ruling in 1997 did not solve the problem; it was a neutral decision, supporting neither side. The court highlighted the illegal suspension of the GNRBS project by Hungary and additionally ruled that the implementation of the new variation by Slovakia was also illegal. As a result, the ICJ determined that the 1977 bilateral treaty was still in effect with the goal of energy production, improved navigation, flood control, regulation of ice removal and environmental protection as stated in the original agreement. Both parties did not accept the decision and the newly elected politicians did not find an agreement, resulting in Slovakia suing Hungary again at the ICJ in 1998, which still did not resolve the problem.

As a result, although the case was never officially solved and declared as such by either conflicting party, both countries have sidelined the GNRBS project in recent years. This is mainly due to changes in foreign policy and government representatives, the NATO and EU accession of Hungary as well as other influences over time. Most scholars agree that there are no real winners in the case of the GNRBS, but the losers are the inland delta of the Danube, the large wetland of Europe and the local inhabitants.[88]

Studies showed that the GNRBS affected fish species due to the restricted and controlled river flow, and the drained flood plains that were seen after October 1992 when the Gabcikovo Dam system and diversion of the water in the bypass canal restricted some fish species from following the water.[89] Additionally, a 40-kilometre stretch of the original channel is badly affected by lack of sufficient water discharge. In some areas surface water and groundwater levels have dropped by up to four metres, greatly affecting the ecological conditions in the area's wetlands. Furthermore, the loss of inland delta also resulted in loss of spawning, feeding and winter grounds for fishes, resulting in a mean annual fish-catch drop of 87 per cent in comparison to the period 1961–1972 (before the construction of the dam system).[90]

## *ICPDR – No Real Role*

The GBNRS showed that there is a need for negotiation and a desire for a common approach to develop joint resources. According to Article 24 of the Danube Convention, disputed parties 'shall seek a solution by negotiation or by any other means of dispute settlement acceptable to the dispute'.[91] If appropriate, the ICPDR can be used as a facilitator between the disputed parties. However, as highlighted above, the ICJ was used as a facilitator as opposed to the ICPDR. As a result, some people questioned why the Danube managing institution was not involved more significantly. However, the GNRBS conflict arose before the establishment of the ICPDR in 1994 and goes back to the bilateral agreement between the two countries in 1977. The ICPDR as a newly established institution did not intervene in the ongoing legal process.

Additionally, the procedure also follows the regulation in the Danube Convention in Article 24, Paragraph 2, which states that if the parties are not able to settle the dispute within 12 months, it should be submitted for a compulsory decision to the ICJ or arbitration in accordance with Annex V of the Danube Convention.[92] Schmeier argues that the ICPDR played an implicit but important role in the process leading up to the ICJ judgement and the implementation of the decision contributing to awareness of sustainable and ecologically cautious infrastructure developments.[93] Others argue that the ICPDR did not have a specific or formal role and supported bilateral negotiations between the riparians.[94] Additionally, the ICPDR provided a forum for cooperative activities between the formerly disputing states, resulting in cooperation over other issues under the ICPDR framework which put the GNBRS in the background.

## *Conclusion: The Gabcikovo–Nagymaros Project*

Changes to the Danube had already been observed prior to the GNRBS system mainly due to dams built in the German and especially the Austrian part of the

Upper Danube reducing the sediment flow and the dredging of the river bed from Slovakia and Hungary.[95] Today, 8,000 hectares of floodplain forests and an extensive network of tributary channels are artificially irrigated. The original river channel only receives 10–20 per cent of the total flow, as the other 80–90 per cent is channelled around the dam towards the power plant, where it is used to generate some 10 per cent of the electricity used in Slovakia. Nevertheless, there are still plans for more dams to be built, for instance on the Bavarian Danube, on the Sava and on the Drava along the Croatian–Hungarian border, where the Novo Virje Dam (planned capacity: 121 MW) would break up the still largely pristine 370-kilometre stretch of river along the Mura and Drava rivers between the Austrian border and the Danube.[96]

The GNRBS showed that there is a need for negotiation and a desire for a common approach to develop joint resources. The ICJ ruled that the principal purposes of the agreement are to uphold electricity production, navigation improvement, flood protection and regional development. Both parties did not accept the decision; however, changes in government representatives and foreign policy sidelined the GNBRS project in recent years. The Gabcikovo–Nagymaros Project technically remains still pending following the decisions on the merits and undetermined negotiations between the parties concerning the implementation of the ICJ ruling. Nevertheless, it is still an important case to show how bilateral problems can develop into challenging situations in river basins. Further, it shows that today, more than 30 years after the beginning of this project, navigation, hydropower and flood prevention are still the key challenges in the Danube River as in several other transnational rivers, including the La Plata River Basin.

## Hazardous Spills – The Baia Mare Incident

Economic development between 1960 and 1989 resulted in the significant worsening of the water quality along the Danube and inland rivers. Since then, water quality has been improving due to two main reasons: the reduction of economic activities and new regulations such as the Polluter Pays Principle. However, water quality is still a key issue as the quality of the levels lie below the 1950 levels. An example of water pollution is the Baia Mare cyanide spill from a gold mine in Romania in 2000 which reached the Tisza River (a major tributary) and then the Danube River, killing large numbers of fish in Hungary and Yugoslavia (see Map 4.4).[97] In total, the cyanide spill affected four riparian states on approximately 2,000 kilometres of the Danube catchment, spilling 50–100 tons of cyanide into the river.[98] The spill has been identified by some as the worst river pollution accident in Europe and as an 'aquatic version of Chernobyl'.[99]

The 2000 Baia Mare cyanide spill was a leak of cyanide from a gold mine near Baia Mare in north-western Romania into the Somes River, later reaching into the Tisza[100] and the Danube rivers, by the gold mining company Aurul SA. Aurul SA is a joint venture of the Australian company Esmeralda Exploration and state-owned Romanian mining concern, REMIN, including private investors

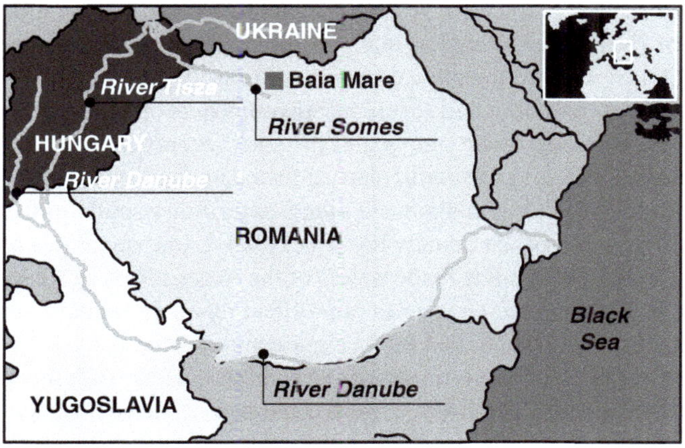


*Map 4.4* Baia Mare incident location.

Source: Adjusted from *BBC News*, 2001.

such as the German-based Dresdner Bank.[101] Aurul SA was established in 1992, obtained environmental permits in 1993, started construction four years later and in 1999 the operational permit was obtained and processing started. When Aurul SA commissioned the plant in 1999, the expected production rate was 1.6 tons of gold and 9.0 tons of silver per year, providing new jobs for 150 Romanians (directly or indirectly employed) and, additionally, 200 jobs during the construction phase in an area with high unemployment.

In general, Aurul SA extracts non-ferrous metals from waste rock piles of mines in the Baia Mare area and uses the cyanide heap-leaching method (extracting gold with cyanide). Most of the mined materials end up in fine slurry, called tailing, which is then stored in tailing dams (storing the hazardous material).[102] In general, the facilities including the tailing dam are closed-circuit systems which avoid discharging emissions into the river system; this was the basis for the EIA of the Aurul SA site. The company was processing and storing solid mining wastes from the extraction of residual gold and silver in a 30-year-old tailing dam when, on the night of 30 January 2000, a dam holding contaminated waters burst and 100,000 cubic metres of cyanide-contaminated water (containing an estimated 100 tonnes of cyanide) spilled over farmland and then into the Somes River.

The flooding washed away a 25-metre-wide and two-and-a-half-metre-deep section of the new tailing dam wall. After the spill, the Somes River had cyanide concentrations over 700 times of the permitted levels. Additionally, heavy metals were also washed into the river, resulting in long-lasting negative impacts on the environment.[103] The pollutants – including cyanide and metal-rich liquid – flowed via different tributaries, first into the Tizsa River and finally through the Danube River into the Black Sea, causing the rapid death of aquatic organisms and animals living in and close to the rivers; around 2,000 kilometres of the Danube basin were affected. After the cyanide entered the Danube, the large volume of river water diluted it, but in some sections it still remained as high as 20 to 50 times above the allowed concentration.[104]

The transboundary pollution had severe negative impacts on biodiversity, the river's ecosystem, drinking water supply and the socio-economic conditions of the local population. Wildlife was particularly affected on the Tisza River where on a stretch virtually all living things were killed, and further south, in the Serbian section, 80 per cent of the aquatic life was killed. Large quantities of fish died due to the toxicity of cyanide in the waters of the rivers, affecting 62 species of fish, of which 20 are protected species (also affecting other animals, such as foxes, otters and ospreys, which died after eating contaminated fish).

The drinking supplies of three countries – Hungary, former Yugoslavia and Romania – were poisoned.[105] Similarly, long-term damage occurred in agricultural products due to the heavy metals in soil and water, affecting not only incomes, but, more importantly, human health due to the chronic effects of pollution.[106] Two years after the spill the ecosystem began to recover, but it was still far from its initial state, as fishermen of Hungary claimed that their catches in 2002 were only one-fifth of their original levels. Biologists argue that it will take around five years to restock the fish and another 10–20 years for river life to return to normal, with some of the damage considered irreversible.[107] Additionally, a flood in April 2000 hit Romania, Hungary and Yugoslavia spreading the sediment of the earlier spill across the Tisza floodplain, endangering soil safety and making it difficult to monitor long-term effects.

The breach in the dam was, according to the United Nations Environmental Program (UNEP) assessment mission, a 'combination of inherent design, deficiencies in the process, unforeseen operation conditions and bad weather'.[108] Others argue that the two main reasons for the disaster include the large volume of rain and melting snow overflowing the tailing dam and construction flaws in the dam.[109] Aurul blamed the flooding on excessive snow falls, and denied responsibility for the spill, arguing that the extent of the poisoning was exaggerated with no proof of linkage.

## *The Role of the ICPDR and Other Conventions*

The Baia Mare incident attracted enormous national and international attention and showed the positive and negative sides of legal regimes, their effectiveness in emergency situations and the compliance of the signatory parties. The Danube Convention includes – similarly to other relevant conventions applicable to the Baia Mare incident – the Polluter Pays and the Precautionary Principle.[110] Other obligations include that the states control the impact of transboundary pollution, prevent hazardous accidents and warn downstream riparian states when a transboundary incident occurs.[111]

As a consequence, the ICPDR has set up an Accident Emergency Warning System (AEWS) of the Danube River Basin, a system of notification which is activated when transboundary water pollution is likely, sending out warning messages to downstream countries when an incident occurs.[112] However, this can be problematic if the incident is not disclosed by the company or government, resulting in delays of action which could worsen the impact and restrict warning systems. Although there was an initial underestimation of the consequences of the spill by the Romanian government, its official response was successful, resulting in timely information exchange with the affected downstream riparians.[113] In the case of the Baia Mare incident, the AEWS mostly worked to warn downstream countries, showing minor possible improvements for the future.

Another obligation of riparian states under the Danube Convention is expressed in Article 6c, which states that countries need to 'minimise the risks of accidental pollution'.[114] Most experts argue that Romania did not fulfil its obligations and tasks under the convention to minimise the risk of accidental pollution by not applying stricter technological standards to the Aurul operations (the operations did not actually breach national Romanian law).[115] However, this is controversial as Article 6c does not specify what 'minimisation of risks' means. Similarly, although Aurul SA complied with Romanian law, the EU found that the EIA was inadequate and did not meet internationally accepted quality standards. The main problems with the EIA included the lack of inclusion of the possibility of severe weather conditions, deficiencies in the understanding of the dam technology in a local context, and the lack of centralised and clear responsibility for the permitting and monitoring process.

Romania sued Aurul SA for the damages of the overflow from the gold mine into the river system. Additionally, negotiations over compensation for damages failed between Hungary and Romania. As a result, Hungarian officials decided to achieve compensation for damages through a civil lawsuit against Esmeralda, the Australian company operating Aurul. The civil court case was filed by Hungary instead of entering another international lawsuit (after the Gabcikovo–Nagymaros case) against a neighbouring country during the country's EU accession period. Hungary and Romania also have a bilateral treaty for security cooperation (the Hungary–Romania Treaty) which incorporates the prevention, reduction and elimination of transboundary pollution.[116]

Overall, the Baia Mare case shows that industrial accident prevention, liability and foreign investments are key areas for further improvements and clearer legislation.[117] Furthermore, the ICPDR's focus was on reducing incidents in the future by mitigation control measures and by prompting countries to provide it with lists of hazardous sites. As highlighted by one interviewee, 'the ICPDR didn't see itself as a legal mechanism for issues – more a forum for consensuses'.[118] The incident also showed that NGOs and public awareness can be critical in putting pressure on national governments regarding environmental hazards, as often local people are unaware of the dangers posed by industrial sites. This is especially relevant in the Romanian context where mining has a long tradition and has a major impact on low water quality. This low awareness can also be seen today in the Mekong River, where local people are unaware of the possible negative impacts from hydropower dams or navigation incidents on fisheries and livelihoods.

### *Conclusion: The Baia Mare Incident*

The northern border region of Romania and Ukraine is rich in gold, silver, lead and copper. It has a long history of mining, containing around 215 tailing dams which are all potential sources of pollution and have constantly resulted in a high level of chronic pollution of the ground, water and air in the region. The Baia Mare incident affected the agriculture and fishery sectors in several countries including Romania, Hungary and Yugoslavia and it remains to be seen if the river ecosystem can be restored to previous levels. The case of the Baia Mare incident shows the problem of transition states adapting environmental law and legislation, especially in terms of implementation, into national law, surveillance and law enforcement.

Additionally, the case also shows the problematic nature of international corporations operating overseas in countries with lower standards and the issue of liability for damage regarding the tailing dam design and the emergency response preparation. This is not unique to the Danube River as similar challenges were seen in the Mekong River case study. Laws and conventions play an important role and Danube riparians have made significant progress. However, the demonstrated case shows there are challenges ahead and improvements needed, especially in transitional countries. Additionally, the Baia Mare incident was the first environmental disaster in which the EU accession featured prominently, as Hungary joined the EU in May 2004 and Romania in 2007.

Further, the incident highlights that the focus needs to shift from damage limitation to prevention of accidents and negative environmental impacts. Additionally, the ICPDR, as well as other active groups such as the WWF, demonstrated that there is a high risk of accidental spills of dangerous chemicals at several mining and pulp-and-paper industries in Eastern Europe.[119] Only five weeks after the Baia Mare cyanide spill another spill of contaminated waters (this time with heavy metals) occurred in the region. A dike burst in Baia Borsa, Romania, and 20,000 cubic metres of zinc-, lead- and copper-contaminated water made its way into the Tisza. In 2001 another cyanide spill happened in Romania, this time through an intentional emptying of cyanide solutions into the Siret River. One of the most recent incidents includes the Ajka red sludge spill in 2010 when a dam broke at an aluminium plant in Hungary releasing 700,000 cubic metres of red sludge, killing ten people and affecting 1,100 hectares of land.[120] There is an urgent need to start abatement activities in the Baia Mare region because of the possibility of future accidents.

## Canalisation/Navigation – the Bystroe Canal

Besides accidental spills and pollution, navigation and related canalisation play an important role in the Danube basin. Until the late 1950s, vessels in Ukraine used the Bystroe and Prorava waterways. However, due to the natural process of waterway silting, both canals closed, impacting on the economic performance of Ukraine. The closure especially affected transport enterprises in the region, which pushed for its own canal within the Ukrainian Danube Delta to connect to the Black Sea. The Bystroe Canal – located in the Danube Delta – was seen as an impractical option in 1997, but was reactivated in 2000 (see Map 4.5).

A feasibility study from 2003 showed that the canal would increase inland shipping by more than 60 per cent, resulting in an additional 4,000 jobs and non-stop two-way traffic.[121] Countries affected by the project are mainly Ukraine, Romania and, to a lesser degree, Moldova. Romania is mainly affected by the social and economic impact on transboundary fishery and migratory fish, but also fears a threat to its economic interests through a better navigation canal in the Ukraine impacting on its canal capacity, utilisation and a possible tax loss due to fewer ships travelling through Romania.

The Ukrainian government started construction of the Bystroe Canal in 2004 in order to boost national economic performance by connecting the country with the commercial waterways in Europe. The Ukraine government argues that the project and the construction of the canal is of economic, geo-strategic and military importance to the country which, amongst others, is reviving its shipping industry and reducing the Romanian monopoly on a navigation canal that increases costs for Ukrainian enterprises.

The Bystroe arm of the Danube River runs through the middle of Ukraine and ends in the Danube Delta which is shared by Ukraine and Romania before flowing into the Black Sea. The German company Josef Mobius Bau AG was used as the general contractor for construction. The construction of the canal was planned in two separate phases, with the first phase resulting in a passage for vessels with a water draught of 5.85 metres expanding to vessels of 7.2 metres in Phase 2. The completion was scheduled for autumn 2005, costing a total of around 32.3 million EUR (with a total cost return in nine years).[122] Widening the canal from 2.5 to 5.85 metres will involve dredging the river bed and deepening it, leading to the dislocation and removal of huge quantities of sand and mud.

Expanding on Mari Koyano's work, the Bystroe Canal project can be divided into six different stages.[123] In Stage 1, from August 2003 to February 2004, Ukraine approved Phase 1 of the project and the joint UNESCO/Ramsar mission raised concerns regarding likely adverse effects. The Danube Convention requested Ukraine to submit the relevant information. In Ukraine itself, several lawsuits were brought to domestic courts concerning the project, but the legal challenges were unsuccessful.[124] During Stage 2, from March 2004 to August 2005, international criticism from different conventions increased due to the start of the execution of the canal; the conclusion was that Ukraine was in breach of its obligations under the Bern Convention.

At the same time, the European Commission started bilateral talks with Ukraine and an expert team of the European Commission and International

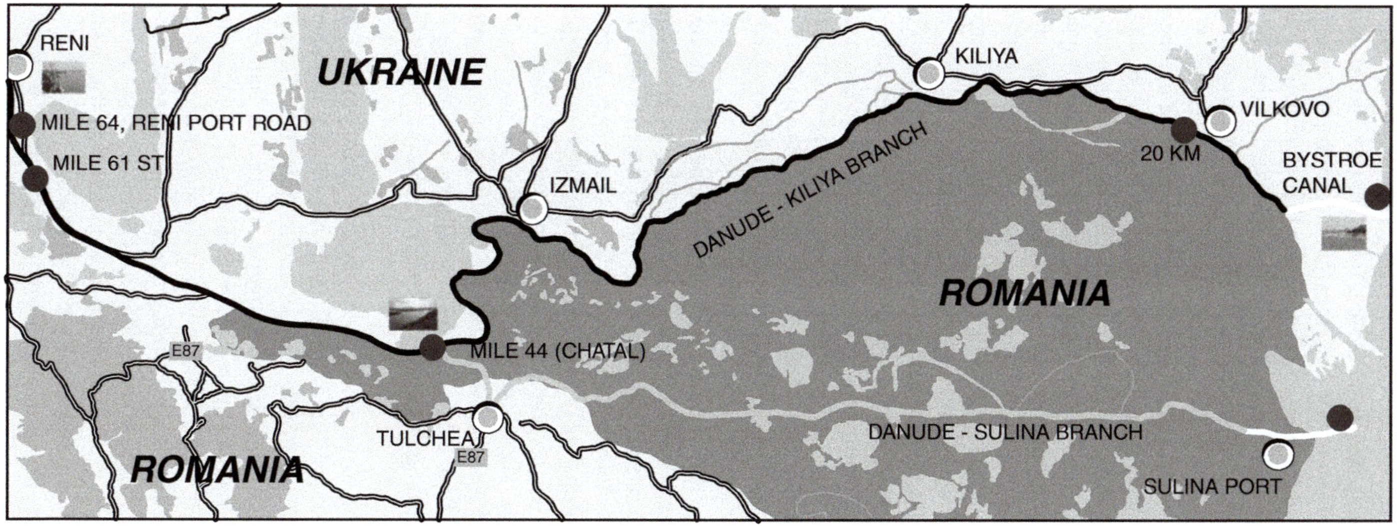


*Map 4.5* The Bystroe Canal project.

Source: Adjusted from Marvel Shipping, 2011.

Convention visited the project in November 2004. The mission concluded that the 'information policy for Phase I of the project was inadequate and the decision-making opaque'.[125] Additionally, Romania took official steps under the Aarhus and Espoo Convention and NGOs became more active.

Following this, in Stage 3 – from September 2005 to October 2006 – Ukraine changed its position and provided more positive responses to international inquiries, for instance reporting to the ICPDR that it would revise the EIA of Phase 2. The United Nations Economic Commission for Europe (UNECE) stated in 2006 that the project had likely significant adverse transboundary impacts in six areas including the loss of floodplain habitat important for fish and birds; habitat loss through dredging; coverage of dump sites; increase of suspended sediment concentration downstream of the dredging site; and increase of turbidity of marine waters.[126] At the same time, initiatives by Ukraine included the suspension of the Bystroe project and the holding of the Odessa Conference on the conservation and sustainable use of the Danube Delta in Ukraine in February 2006, which was seen as a success.[127] Additionally, a framework for preparing a river basin management plan for the Delta was started under the ICPDR framework.

However, the approach of the Ukrainian government changed again (mainly due to a political change in the government and the increased power of the transport department) between November 2006 and April 2008 (Stage 4) when Ukraine became less cooperative towards the international community and resumed the Bystroe project. This included new dredging (as it was silted up past its previous depth) and the re-opening of the canal in April 2007, with an official opening of the complete deep-water shipping canal by 2009.[128] Ongoing maintenance work necessary to eliminate the sediments in order to keep the canal navigable is very cost intensive (which is also the case in other navigational projects, for instance along the Upper Mekong). The official opening resulted in formal actions by the different international conventions stating, for example, that under the Espoo convention the non-compliance of Ukraine was decided by the Meetings of Parties in May 2008. Additionally, the UNECE stated in early 2008 that the project will have 'significant adverse transboundary effects' on the environment and should be stopped until proper evaluation has been undertaken.

This viewpoint was supported by a meeting of intergovernmental organisations in May 2008, including the ICPDR, the Bern Convention and UNESCO's Man and Biosphere Program, stating in regard to the Bystroe project that,

> commitments and obligations of Ukraine under international conventions have not been adhered to in the development of the project and that the December 2007 decision by Ukraine to implement the project in full – if done without implementing these obligations – would be contradictory under various international conventions and agreements.[129]

This also included the EIA of Phase 2 which was declared invalid by the Ukrainian court due to the breach of the Espoo Convention and other international agreements. However, Ukraine proceeded with the project arguing that all international obligations had been fulfilled.

Since May 2008, Stage 5, a 'Strategy on the Implementation of Provisions of the Espoo Convention', has been prepared by the Ukraine government, outlining the future of the project and future cooperation with neighbouring riparians. Later in 2008 the project changed focus again with the goal of finding a common ground for international cooperation and open dialogue regarding the Bystroe Canal after discussion on the project had stalled earlier. Ukraine expressed that it would participate in an active dialogue with the appropriate bodies and that it was ready to reconsider its decision on the project.[130] This was supported by a repeal of the final decision regarding Phase II of the project, which would not start until appropriate procedures were implemented.[131] In Stage 6, from 2010 until today, Ukraine decided to start work on the full-scale implementation of the Danube–Black Sea Navigation Route. Dissatisfaction arose again in September 2011 due to a lack of communication from Ukraine and further changes in 2012 which increased opposition by Romania and Moldova. Political instability in Ukraine at the end of 2014, which is still ongoing, impacted further on the project, with unclear outcomes and ongoing concerns by affected neighbours and international organisations.

As highlighted above, opposition to the project started from several areas focusing on the infringement of signed international conventions, the endangering of wildlife, the loss of work for the fishing industry and transboundary impacts. Key negative environmental impacts include the triggering of a more active river flow process affecting the ecological balance of the area, including habitat of plants and animals; the pollution caused by petrol and other toxic substances; the creation of transboundary effects on the Romanian side; and the serious impacts on socio-economic conditions of the local population.[132]

One of the most controversial points of the project was the location of the canal as it cuts through the Danube Delta Biosphere Reserve protected under the UNESCO Man and Biosphere Program, part of the Danube Delta, which is home to over 300 species of birds and 45 varieties of fish. Besides the impact on fisheries, the Bystroe Canal has impacted on the water quality and biodiversity existing in the Transboundary Biosphere Reserve designated by UNESCO. In 1998, Ukraine and Romania recorded 1,689 plant species and 3,460 animal species, including unique migratory birds. Additionally, the Bystroe Canal is one of two remaining migration routes and possibly the most significant because sturgeons and Danube herring migrate from the north-western part of the Black Sea. Additionally, it is the migration site for 95 per cent of the Danube sardine population.[133]

### *The ICPDR and Its Function*

The ICPDR has been involved in developments of the navigation project since 2003 and has established a dialogue with the representatives of the Ukraine and neighbouring countries. Former ICPDR President Catharine Day raised early concerns about the possible transboundary environmental impacts from the construction of the canal and expressed these concerns at the December 2003 Ordinary Meeting. A day later it was called upon the Ukrainian government to halt the future states of the construction until a proper international evaluation of the environmental consequences of the project can be undertaken.[134] This was supported by the ICPDR Executive Secretary Phillip Weller who wrote to an NGO that the 'ICPDR is very concerned about this project and the possible environmental effects it may have both in the Danube River and Black Sea ecosystem' and especially in regards to the 'require[d] exchange of information on planned activities, which for reason of their character, are likely to cause transboundary impact'.[135]

Regular reports have been presented to the Ordinary Meeting and Standing Groups of the ICPDR from 2004 by the Ukraine as requested under Article 10 of the Danube Convention. This article requires that projects with transboundary relevance should be presented to the other contracting parties.[136] The key focus of the ICPDR was to reinforce the need to fulfil the requirements of the Espoo Convention, to inform neighbouring countries of the project and to have input to develop or adjust the project to eliminate transboundary effects. However, although Ukraine has been attending ordinary meetings of the ICPDR since September 2004, sufficient information was often not available. Similar to the other international organisations, the ICPDR kept adopting resolutions which requested the Ukraine to submit further information, take consultations and provide an appropriate EIA.

In the following years, the ICPDR facilitated different initiatives, including the sub-basin management plan of the Danube Delta, and a trilateral Memorandum of Understanding (MoU) was signed at the end of 2007. Other projects included the promotion of integrated planning of navigation and environmental improvements projects in the Danube River and its basins.[137] Additionally, the ICPDR requested further information regarding the adoption of an appropriate EIA. Another shift included the move away from conflict between Ukraine and Romania towards a cooperative approach of both parties for a sustainable development of the Danube Delta. As highlighted by an expert, it was important to keep Ukraine involved, especially the Ministry of Foreign Affairs which would bring information back to the country.

Additionally, the major criticism of the Ukraine government focused more on non-compliance with existing, signed conventions and the involvement of the public in the EIA than with the project itself. The active role of the ICPDR resulted in an increased credibility of the institution as a place for the intelligent discussion of issues and exchange of information. However, as noted by one expert, this was not easy to achieve because there was opposition from within the riparian states.[138] The main concern of some riparians was to set a precedent where the ICPDR would get involved in bilateral issues, as it was unclear at the beginning if the project was a bilateral or a basin-wide problem. Yet, the project was seen as a basin-wide threat and as a test case for the ICPDR on whether it was able to stand up for the environment and the basin. Today, a major problem is the lack of capacity and financing to get involved further in the issue which was scaled down in recent years.

## *Conclusion: The Bystroe Canal Project*

The Bystroe Canal shows that environmental interests still come second to economic, strategic-military and political interests, even in the case of several existing bilateral and multilateral agreements and international conventions signed in a strong environmental framework (although Ukraine is not part of the EU it is included in the European Neighbourhood Policy). Regarding the Bystroe Canal, the function of the ICPDR moved from an observer role in the early stages to a mediating role. Overall, the ICPDR still played an important role in the negotiation process between both parties and the other international players. One of the key attributes of the ICPDR is the focus on sub-basin development and the sustainable development of the Danube Delta. Overall, as one interviewee observed, the Bystroe Canal was the most successful achievement of the ICPDR in comparison to the other two cases (Gabcikovo–Nagymaros and Baia Mare) and conferred credibility on the institution.[139]

Regarding the EU framework, there are several positive and negative factors which influenced the project and the process dealing with it. On the positive side, there is the increase of multilateral environmental agreements and the corresponding regional institutions, the significance of procedural regulations and prior consultations, the EIA, information exchange and institutionalisation in the EU context. Additionally, the involvement of NGOs in the formal process, the emphasis on public participation, and the bilateral exchange between the European Commission and Ukraine were drivers behind managing the Danube Delta conflict. On the negative side, one of the biggest threats today are EU development and plans as part of the EU Trans-European Transport Network (TEN-T) to dike and dredge the Danube further to allow for larger ships. One of the key criticisms by local and international NGOs is the lack of integrated planning and linked decision-making in regards to inland navigation on the lower Danube.[140] This is an internal problem of the EU and its funding, and can be highlighted by the contrasting elements of the EU WFD and the EU TEN-T.[141]

Moreover, the case shows that key challenges facing Ukraine regarding compliance are not unique and can be found in several of the other micro-level case studies, including the lack of will of the state/states concerned, uncertainty about the content of international agreements, shortage of resources and conflict with domestic legal and administrative systems. These are still visible in the case of the Bystroe Canal, although there have been some improvements in recent years. Additional challenges and future problems regarding the project include possible ship accidents or industrial disasters if the water is affected by oil, chemical products, fertilisers and other hazardous goods regularly transported on inland waterways.[142] For instance, between 1983 and 2003, 455 ship accidents occurred on the Danube, with 30 of them resulting in significant water pollution.[143] One recent example is the oil spill by a sunken barge in 2011 in Bulgaria's largest port, Ruse, which threatened to leak into the main river with possible disastrous impacts on the environment.[144]

## Summary and Conclusion

The Danube River is a case study which, at first blush, appears completely different to the other two case studies. First, economic and political developments in Europe are different to those in Southeast Asia and South America as they are dominated by a strong EU framework and the ambitions of Eastern European countries to join the EU, which means that Europe is more institutionalised than the other two regions. Second, the riparian countries in the Danube are more homogenous than those in the other two case studies, with similar political systems and cooperative dimensions. However, if one looks more closely at the case study and its close linkages between Western and Eastern Europe, it becomes clear that some of the riparian countries are less developed and continue to confront the challenges of corruption and the establishment of reliable legislation. Also, as in the Mekong countries, the Danube countries were ideologically divided during the Cold War, a situation that still has implications today. Further, although cooperation on transboundary rivers such as the Danube works relatively well, there are still problems, including the incorporation of the less developed countries with low financial capacities, as well as challenges from climate change which will affect the Danube as much as the other two river basins.

The Danube case study further shows that it is important to understand Europe's historical ties, the developments of the EU framework and the upstream–downstream relations in order to get a full picture of water resource management along the Danube and its tributaries. Laws and conventions play a major role and Danube riparians are quite advanced. Yet, the demonstrated cases show that there are still problems and challenges and that improvements are necessary, especially in transitional countries. The three selected case studies – the Gabcikovo–Nagymaros hydropower development, the Baia Mare cyanide spill and the Bystroe Canal project – show that the EU framework has advantages in environmental standards and developments; however, there are still gaps which need to be filled, including the problematic requirement of having to adjust national legislation to international standards, which can often take several years. Also, occasionally EU projects can be contradictory as in the case of the Bystroe Canal when TEN-T infrastructure projects conflict with the WFD. However, this case study also supports findings that EU accession and political unification of the region are critical for cooperation, which is also driven by a desire of mid- and downstream countries to achieve similar economic standards to the upstream countries Germany and Austria.

Additionally, in a similar fashion to the other previously discussed key decisions in the Mekong, natural resource management and river cooperation are used as a political instrument and a show of power – for instance, after the Gabcikovo withdrawal of Hungary, which was linked to the revolution in the country. The ICPDR, although limited, has had a positive effect on the riparians. However, as highlighted above, the power relations amongst the countries and their historical background are still important and have influence on contemporary developments. It is also possible that the ICPDR will face challenges in the future when the previously less dominant states become more advanced and will demand to influence the agenda. Different levels of involvement of the ICPDR in the three cases are evident, from nearly no role in the Gabcikovo project, to a mediating role in the Bystroe project (which is also seen as the most successful of the three cases in regards to the reaction and work of the ICPDR). Similarly, the role of NGOs and public participation in the Baia Mare and the Bystroe Canal project is increasingly important and can influence countries' decisions and provide awareness within the community; however, these benefits are still limited in some of the less developed countries in Eastern Europe and can certainly be improved in the future.

Overall, the Danube case study shows that although the geographical positioning is uneven with the most developed countries located upstream and the least developed downstream, due to the strong commitment by the upstream countries, Germany and Austria, and other regional incentives (mainly EU accession), cooperation is working well in the ICPDR; however, this might change in the future as the previously less developed countries become 'visible' and could build a block to influence the agenda and direction of the organisation.

## Notes

1 Domokos Miklos and Jeno Sass, "The Danube River and its Basin Physical Characteristics, Water Regime and Water Balance," in *Hydrological Processes of the Danube River Basin*, ed. Mitja Brilly (Dordrecht: Springer Science and Business Media, 2010), 26.
2 Domokos Miklos, "History and Results of the Hydrological Co-Operation of the Countries Sharing the Danube Catchment (1971–2008)," in *Hydrological Processes of the Danube River Basin*, ed. Mitja Brilly (Dordrecht: Springer Science and Business Media, 2010), 3–4.
3 Ibid.; Pavel Petrovic, Katarina Mravcova, Ladislav Holko, Zdenek Kostka, and Pavol Miklanek, "Basin-Wide Water Balance in the Danube River Basin," in *Hydrological Processes of the Danube River Basin*, ed. Mitja Brilly (Dordrecht: Springer Science and Business Media, 2010), 683.
4 International Commission for the Protection of the Danube River (ICPDR), "Danube Facts and Figures," ICPDR, Vienna, 2010a, 13.
5 Stephen McCaffrey, "The Danube River Basin," in *The Multi-Governance of Water*, eds. Matthias Finger, Ludivine Tamiotti, and Jeremy Allouche (New York: State University of New York Press, 2006), 81–2.
6 Svetlana Mihic, Mirjana Golusin, and Milan Mihajlovic, "Policy and Promotion of Sustainable Inland Waterway Transport in Europe – Danube River," *Renewable and Sustainable Energy Reviews* 15 (2011): 1801–9.
7 ICPDR, "Dams & Structures," www.icpdr.org/main/issues/dams-structures (accessed 13.11.2013).

8 Ibid.
9 Miklos and Sass, "The Danube River and its Basin Physical Characteristics, Water Regime and Water Balance."
10 Miklos, "History and Results of the Hydrological Co-Operation of the Countries Sharing the Danube Catchment (1971–2008)."
11 Frederick D. Gordon, "The 1994 Convention on Cooperation for the Protection and Sustainable Use of the River Danube (High Level of Cooperation)," in *Freshwater Resources and Interstate Cooperation*, ed. Frederick D. Gordon (Albany, NY: State University of New York, 2008), 100.
12 Ibid.
13 Brian Erskine, "Disaster on the Danube," in *Borders and Border Politics in a Globalizing World*, eds. Paul Ganster and David E. Lorey (Oxford: Rowman and Littlefield, 2005), 238; ICPDR, "Danube Facts and Figures."
14 Ibid.
15 ICPDR, "Danube Facts and Figures – Austria," ICPDR, Vienna, 2006a.
16 ICPDR, "Dams & Structures."
17 Telephone interview with a former Chief Executive Officer (CEO) of the ICPDR on the 20.12.2013.
18 ICPDR, "Danube Facts and Figures – The Czech Republic," ICPDR, Vienna, 2007d.
19 Ibid.
20 Ibid., 5.
21 ICPDR, "Danube Facts and Figures – The Slovak Republic," ICPDR, Vienna, 2007e.
22 ICPDR, "Danube Facts and Figures – Hungary," ICPDR, Vienna, 2006b.
23 Ibid.
24 European Commission, "EU Strategy for the Danube Region – First Contribution by Slovenia," European Commission, Brussels, 2010b.
25 ICPDR, "Danube Facts and Figures – Croatia," ICPDR, Vienna, 2010b.
26 Ibid.
27 Ibid., 3.
28 European Commission, "Danube Facts and Figures – Montenegro," ICPDR, Vienna, 2010c.
29 European Commission, "Danube Facts and Figures – Romania," ICPDR, Vienna, 2006c.
30 National Audit Office Bulgaria, "Report on the Regularity and Performance Audit of the Application of the Provisions of the Danube River Protection Convention," National Audit Office Bulgaria, Sofia, Bulgaria, 2001, 7–8.
31 ICPDR, "Danube Facts and Figures – Republic of Moldova," ICPDR, Vienna, 2007c.
32 Ibid., 3.
33 Silviu Costachie, Ionica Soare, and Daniel Dieaconu, "The Nature of State Trans-border Reterritorialization in the Lower Danube Lands," *Geopolitics, History, and International Relations* 2, no. 1 (2010): 152–7.
34 John Pinder and Simon Usherwood, *The European Union – a Very Short Introduction*, 2nd ed. (Oxford: Oxford University Press, 2007).
35 The other new EU countries not in the Danube Basin are Estonia, Cyprus, Latvia, Lithuania, Malta and Poland.
36 European Commission, "Enlargement from 6 to 27 Members and Beyond," http://ec.europa.eu/enlargement/archives/pdf/press_corner/publications/corpus_en.pdf (accessed 04.06.2012).
37 United Nations Environmental Program (UNEP)/GRID–Arendal, "Balkan Vital Graphics – Environment without Borders," UNEP/GRID–Arendal, Zemun, Serbia, 2007.

38 Ibid.

39 John McCormick, *Understanding the European Union – A Concise Introduction*, 3rd ed. (Hampshire: Palgrave Macmillan, 2005), 202; Miranda Schreurs, "Environmental Protection in an Expanding European Community: Lessons from Past Accessions," *Environmental Politics* 13, no. 1 (2004): 27.

40 Pinder and Usherwood, *The European Union – A Very Short Introduction*.

41 Pinder and Usherwood, *The European Union – A Very Short Introduction*; Schreurs, "Environmental Protection in an Expanding European Community: Lessons from Past Accessions."

42 European Commission, "Enlargement," http://ec.europa.eu/environment/enlarg/index_en.htm (accessed 04.06.2012).

43 Interview with an EU-Enlargement expert from the European Commission in Brussels, Belgium on the 05.06.2012.

44 Interview with an EU-Enlargement expert from the European Commission in Brussels, Belgium on the 05.06.2012; Joann Carmin and Stacy D. Vandeveer, "Enlarging EU Environments: Central and Eastern Europe from Transition to Accession," *Environmental Politics* 13, no. 1 (2004): 8; John M. Kramer, "EU Enlargement and the Environment: Six Challenges," *Environmental Politics* 13, no. 1 (2004): 295.

45 European Commission, "Directive 2000/60/EC of the European Parliament and of the Council of 23 October 2000 Establishing a Framework for Community Action in the Field of Water Policy," *Official Journal of the European Communities* L327 (2000), 1.

46 European Commission, "Directive 2000/60/EC of the European Parliament and of the Council of 23 October 2000 Establishing a Framework for Community Action in the Field of Water Policy."

47 European Commission, "Introduction to the New EU Water Framework Directive," http://ec.europa.eu/environment/water/water-framework/info/intro_en.htm (accessed 15.06.2012).

48 McCaffrey, "The Danube River Basin."

49 Stephen Stec, Alexios Antypas, Tonya Jansen, and Eszter Gulacsy, "Transboundary Environmental Governance and the Baia Mare Cyanide Spill," *Review of Central and East European Law* 4 (2001): 660–1.

50 Later called Danube Convention.

51 Sovereign states with a considerable part of hydrological catchment area of the Danube River, which is defined as a share being bigger than 2000 square kilometres are called Danubian States (ICPDR, 1994).

52 ICPDR, "ICPDR – Contracting Parties," www.icpdr.org/icpdr-pages/contracting_parties.htm (accessed 11.01.2012).

53 ICPDR, "Convention on Cooperation for the Protection and Sustainable Use of the Danube River (Danube River Protection Convention)," ICPDR, Sofia, Bulgaria, 1994, Articles 2, 3.

54 ICPDR, "Permanent Secretariat of the ICPDR," www.icpdr.org/icpdr-pages/secretariat.htm (accessed 04.06.2012).

55 Andrea Haefner, "Cooperation and Conflict Along Transnational Rivers: The Role of Regional Institutions," in *Presentation at the Oceanic Conference for International Studies* (Melbourne, July, 2014), 8; ICPDR, "Permanent Secretariat of the ICPDR."

56 Also called 'DANUBIS'.

57 ICPDR, "Permanent Secretariat of the ICPDR."

58 Interview with an ICPDR senior expert on the 26.4.2012 in Vienna, Austria.

59 Interview with an ICPDR senior expert on the 26.4.2012 in Vienna, Austria.

60 Interview with a senior expert working on the Danube and the ICPDR for several years on the 25.09.2013 in Brisbane, Australia.

61 Interview with an ICPDR senior expert on the 26.4.2012 in Vienna, Austria.
62 Interview with a senior expert working on the Danube and the ICPDR for several years on the 25.09.2013 in Brisbane, Australia.
63 ICPDR, "ICPDR – About Us," www.icpdr.org/icpdr-pages/about_us.htm (accessed 04.06.2012).
64 Telephone interview with a former CEO of the ICPDR on the 20.12.2013.
65 Jerome Delli Priscoli and Aaron T. Wolf, *Managing and Transforming Water Conflicts* (Cambridge: Cambridge University Press, 2009), 151.
66 Danube Commission, "Welcome to the Danube Commission Website," www.danubecommission.org/index.php/en_US/welcome (accessed 18.06.2015).
67 Danube Commission, "Welcome to the Danube Commission Website."
68 Danube Commission, "Danube Commission – Officials of the Secretariat," www.danubecommission.org/index.php/en_US/secretariat (accessed 18.06.2015).
69 European Commission, "Danube Strategy: Transport Ministers to Cooperate on Navigability," http://europa.eu/rapid/pressReleasesAction.do?reference=IP/12/579&format=HTML&aged=0&language=EN (accessed 12.06.2012).
70 European Commission, "Danube Strategy: Transport Ministers to Cooperate on Navigability."
71 Miklos and Sass, "The Danube River and its Basin Physical Characteristics, Water Regime and Water Balance."
72 In 1971 the Danube River only included eight countries, namely Germany, Austria, Czechoslovakia, Hungary, Yugoslavia, Romania, Bulgaria and the Soviet Union. United Nations Educational, Scientific and Cultural Organisation (UNESCO), "IHP Danube Cooperation," www.unesco.org/new/en/venice/natural-sciences/water/danube-cooperation/ (accessed 07.06.2012); Miklos, "History and Results of the Hydrological Co-Operation of the Countries Sharing the Danube Catchment (1971–2008)," 5.
73 UNESCO, "IHP Danube Cooperation."
74 UNESCO, "Principles," www.unesco.org/new/fileadmin/MULTIMEDIA/FIELD/Venice/pdf/NS/Principles.pdf (accessed 07.06.2012).
75 UNESCO, "IHP Danube Cooperation."
76 Interview with a WWF expert on the Danube River in Vienna, Austria on the 24.05.2012.
77 International Association of Danube Research (IAD), "International Association for Danube Research – Home," www.iad.gs/index.php?item=home (accessed 22.06.2012).
78 European Anglers Alliance (EAA), "European Anglers Alliance – History," www.eaa-europe.org/about-eaa/about-eaa.html (accessed 22.06.2012).
79 Interview with a WWF expert on the Danube River in Vienna, Austria on the 24.05.2012.
80 Interview with an EU-Enlargement expert from the European Commission in Brussels, Belgium on the 05.06.2012.
81 United Nations Watercourses Convention, "Article 33," http://legal.un.org/ilc/texts/instruments/english/conventions/8_3_1997.pdf (accessed 01.03.2014), 12; Stephen Deets, "Solving the Gabcikovo–Nagymaros Dam Conflict," Wilson Center, Washington, DC, 1998.
82 Scott E. Smith, Ferenc Szilagyi, and Lajos Horvath, "Environmental Impacts of the Gabcikovo Barrage System to the Szigetkoez Region," *Clean Technologies and Environmental Policy* 4, no. 1 (2002): 54–64.
83 Deets, "Solving the Gabcikovo–Nagymaros Dam Conflict," 1–2.
84 David Mitchell, "Proposed Danube Dam at Hainburg: WWF Protest and Upheld Injunction," *Environmental Conservation* 12, no. 1 (1985): 81; Erskine, "Disaster on the Danube."
85 Erskine, "Disaster on the Danube."

86 Victor Pochat, "Dams and Development Project: Identification, Collection of Information and Compilation of Relevant Practices Concerning the Integration into Policy/Normative Frameworks and Implementation of Key Priority Issue: International Policy in Shared River Basins," www.unep.org/dams/files/Compendium/Report_InterRivers.pdf (accessed 10.10.2014).

87 During that time air pollution was a major problem in the big Slovakian cities and affected millions of people.

88 Eugene K. Balon and Juraj Holcik, "Gabcikovo River Barrage System: The Ecological Disaster and Economic Calamity for the Inland Delta of the Middle Danube," *Environmental Biology of Fishes* 54, no. 1 (1999): 11; Pochat, "Dams and Development Project: Identification, Collection of Information and Compilation of Relevant Practices Concerning the Integration into Policy/Normative Frameworks and Implementation of Key Priority Issue: International Policy in Shared River Basins."

89 Gabor Guti, "Ecological Impacts of the Gabcikovo River Barrage System with Special Reference to Umbra Krameri Walbaum, 1792, in the Szigetköz Floodplain," in *Annalen des Naturhistorischen Museum Wien, Serie B fuer Botanik und Zoologie [Annals of the Museum of Natural History Vienna, Series B for Botany and Zoology]* (Vienna: Naturhistorisches Museum Wien [Museum of Natural History Vienna], 1995), 467.

90 Balon and Holcik, "Gabcikovo River Barrage System: The Ecological Disaster and Economic Calamity for the Inland Delta of the Middle Danube," 5.

91 ICPDR, "Convention on Cooperation for the Protection and Sustainable Use of the Danube River (Danube River Protection Convention)," Article 24.

92 Ibid., Article 24, Paragraph 2.

93 Susanne Schmeier, "The Danube River Basin and the ICPDR: Strong Achievements by a Narrow Institution," in *Governing International Watercourses – River Basin Organizations and the Sustainable Governance of Internationally Shared Rivers and Lakes*. (Abingdon: Routledge, 2013c), 209.

94 Telephone interview with a former CEO of the ICPDR on the 20.12.2013.

95 Balon and Holcik, "Gabcikovo River Barrage System: The Ecological Disaster and Economic Calamity for the Inland Delta of the Middle Danube."

96 Ibid.

97 Former Yugoslavia nowadays includes Serbia and Montenegro. *BBC News*, "One Year On: Romania's Cyanide Spill," *BBC News*, 31.01.2001.

98 Chris Hudson, "The Role of International Environmental Law in the Protection of the Danube River Basin: The Baia Mare Cyanide Spill," *Colorado Journal of International Environmental Law and Policy* 12, no. 1 (2001): 16; UNEP/UN Office for the Co-ordination of Humanitarian Affairs (OCHA), "UNEP/OCHA Assessment Mission – Cyanide Spill at Baia Mare Report," ed. UNEP/OCHA, Geneva, 2000.

99 Emma Batha, "Death of a River," *BBC News Online*, 15.02.2000; *BBC News*, "Pollution Is a Worldwide Problem Which Does not Respect National Boundaries and Is Likely to Intensify as the Spread of Industrial Development Continues," *BBC News*, 13.12.2004.

100 The Tisza River is the longest tributary of the Danube and the largest catchment area. ICPDR, "The Danube River Basin – Facts and Figures," 6.

101 See Stec *et al.*, "Transboundary Environmental Governance and the Baia Mare Cyanide Spill," footnote 9, 644.

102 Hudson, "The Role of International Environmental Law in the Protection of the Danube River Basin: The Baia Mare Cyanide Spill," 370.

103 Cyanides are known for their immediate impact on the ecosystem and fisheries, whereas heavy metals act acutely and chronically, having longer-lasting impact and could persist in the affected environment (especially in sediments). Premysl Soldan, Michal Pavonic, Jan Boucek, and Jiri Kokes, "Baia Mare Accident – Brief

Ecotoxiological Report of Czech Experts," *Ecotoxicology and Environmental Safety* 49, no. 3 (2001): 256–7; Krista Harper, "'Wild Capitalism' and 'Ecocolonialism': A Tale of Two Rivers," *American Anthropologist* 107, no. 2 (2005): 223.

104 Gavin Davids, "The Baia Mare Cyanide Spill," http://news.in.msn.com/gallery.aspx?cp-documentid=3460600&page=7 (accessed 06.01.2014).

105 Ibid.; Harper, "'Wild Capitalism' and 'Ecocolonialism': A Tale of Two Rivers"; UNEP/OCHA, "UNEP/OCHA Assessment Mission – Cyanide Spill at Baia Mare Report."

106 Stec *et al.*, "Transboundary Environmental Governance and the Baia Mare Cyanide Spill," 647.

107 Batha, "Death of a River."

108 Following requests from the Governments of Hungary, Romania and the Federal Republic of Yugoslavia, and consultations with European Environment Commissioner Margot Wallström and the UN Office for the Co-ordination of Humanitarian Affairs (OCHA), Klaus Toepfer, the Executive Director of the United Nations Environment Programme (UNEP) announced on 18 February 2000 that a team of international experts would be sent to the affected area to carry out a scientific analysis of the environmental damage caused by the spill. The mission was a joint venture of UNEP and OCHA, organised by the Joint UNEP/OCHA Environment Unit, and headed by the Director of UNEP's Regional Office for Europe. UNEP/OCHA, "UNEP/OCHA Assessment Mission – Cyanide Spill at Baia Mare Report," 44.

109 Stec *et al.*, "Transboundary Environmental Governance and the Baia Mare Cyanide Spill," 644; Andreas Bernstorff and Judit Kanthak, "The Real Face of the Kangaroo," Greenpeace, Amsterdam, 2000.

110 ICPDR, "Convention on Cooperation for the Protection and Sustainable Use of the Danube River (Danube River Protection Convention)," Article 2, Paragraph 4.

111 Ibid.

112 Ibid., Article 16; ICPDR, "The Danube River Basin – Facts and Figures."

113 Stec *et al.*, "Transboundary Environmental Governance and the Baia Mare Cyanide Spill."

114 ICPDR, "Convention on Cooperation for the Protection and Sustainable Use of the Danube River (Danube River Protection Convention)," Article 6c.

115 UNEP/OCHA, "UNEP/OCHA Assessment Mission – Cyanide Spill at Baia Mare Report," 11.

116 Stec *et al.*, "Transboundary Environmental Governance and the Baia Mare Cyanide Spill."

117 The Sandoz incident in Switzerland in 1986 resulted in the worst pollution of the Rhine River in decades; however, many aspects raised after the incident had not been resolved at the time of the Baia Mare incident, which shows that there is an immediate response to accidents. However, the long-term changes are not as extensive as predicted, especially those regarding problems of liability. Ibid.

118 Telephone interview with a former CEO of the ICPDR on the 20.12.2013.

119 ICPDR, "Regional Inventory of Potential Accidental Risk Spots in the Tisa Catchment Area of Romania, Hungary, Ukraine and Slovakia," ICPDR in cooperation with Zinke Environment Consulting for Central and Eastern Europe, Vienna, 2000.

120 ICPDR, "The Danube River Basin – Facts and Figures," 22; Hungarian Ministry of Foreign Affairs, "Information on the Accidental Pollution and Related Mitigation Measures of the 'Red Mud Spill at Ajka'," Hungarian Ministry of Foreign Affairs, Budapest, 2010.

121 ICPDR, "The Bystroe Canal: Dredging up Many Questions," ICPDR, Vienna, 2004a; Kirstie Shepherd, "Finding a Common Ground for International Cooperation," *Danube Watch*, 2/2008.

122 ICPDR, "The Bystroe Canal: Dredging up Many Questions."

123 Mari Koyano, "Effective Implementation of International Environmental Agreements: Learning Lessons from the Danube Delta," in *Public Interest Rules of International Law*, eds. Teruo Komori and Karel Wellens (Farnham: Ashgate Publishing Group, 2009), 261–3; Mari Koyano, "The Significance of the Convention on Environmental Impact Assessment in a Transboundary Context (Espoo Convention) in International Environmental Law: Examining the Implications of the Danube Delta Case," *Impact Assessment and Project Appraisal* 26, no. 4 (2008): 299–314.

124 Koyano, "Effective Implementation of International Environmental Agreements: Learning Lessons from the Danube Delta."

125 Ibid., 9.

126 United Nations Economic Commission for Europe (UNECE), "Bystroe Canal Project and Convention on Environmental Impact Assessment in a Transboundary Context (Espoo, 1991)," www.osce.org/eea/32889?download=true (accessed 01.11.2014), 1–14.

127 Koyano, "Effective Implementation of International Environmental Agreements: Learning Lessons from the Danube Delta."

128 Kirstie Shepherd, "Work Resumes on Bystroe Canal," *Danube Watch*, 3/2006; World Wide Fund for Nature (WWF), "Bystroe Canal/Danube–Black Sea Deep Navigation Channel," WWF, Vienna, 2009a.

129 Ramsar Convention on Wetlands, "Bystroe Canal Project under International Scrutiny," www.ramsar.org/news/bystroe-canal-project-under-international-scrutiny (accessed 13.04.2014).

130 Ibid.

131 Council of Europe, "Convention on the Conservation of European Wildlife and Natural Habitats – Application of the Convention – Summary of Case Files and Complaints," in *Standing Committee* (Strasbourg, 27–30 November, 2012).

132 Mircea Dutu, "Legal Implications of the 'Bystroe' Danube–Black Sea Canal Project," Ecological University of Bucharest, Bucharest, n.d., 4.

133 Ibid., 5–6.

134 As stated in the ICPDR Statement on the opening of the Danube–Black Sea deep navigation canal (Bystroe area) in Ukraine. ICPDR, "The Bystroe Canal: Dredging up Many Questions."

135 ICPDR, "Re: Letter Expressing Concerns About Ukraine Violation of the Convention on Cooperation for the Protection and Sustainable Use of the Danube River," ICPDR, Vienna, 2004b.

136 Ramsar Convention on Wetlands, "Bystroe Canal Project under International Scrutiny."

137 Koyano, "Effective Implementation of International Environmental Agreements: Learning Lessons from the Danube Delta," 267.

138 Telephone interview with a former CEO of the ICPDR on the 20.12.2013.

139 Telephone interview with a former CEO of the ICPDR on the 20.12.2013.

140 WWF, "Save the Danube as a Lifeline! – Steps Towards Sustainable Navigation. Common NGO Position on Navigation in the Danube Basin," WWF, Vienna, 2009b, 5.

141 Ibid.

142 WWF, "Save the Danube as a Lifeline! – Steps Towards Sustainable Navigation."

143 ICPDR, "Danube Facts and Figures – Romania," 4.

144 *Phys.org*, "Sunken Barge Threatens Oil Spill in Danube, WWF Says," http://phys.org/news/2012-11-sunken-barge-threatens-oil-danube.html (accessed 20.11.2013); *Environmental News Service*, "Small Oil Leak Points up Big Danube River Issues," http://ens-newswire.com/2012/11/27/small-oil-leak-points-up-big-danube-river-issues/ (accessed 21.10.2013).

# 5 La Plata River Basin: Overlapping Bilateral and Multilateral Treaties

The last case study of this book analyses the La Plata River in South America, the fifth largest river in the world and one of the most threatened due to new infrastructure, hydrological alternations for shipping, proposed large dams and climate change. The La Plata River flows through one-fifth of the South American continent and covers the riparians Brazil, Paraguay, Uruguay, Bolivia and Argentina. As a result of its geographical location, developments and cooperation are closely linked to the past and current Brazil–Argentina relationship, various existing treaties and regional developments including the Southern Cone Common Market (Mercosur). This case study highlights the political, economic and ecological importance of the La Plata region, as well as the significance of future regional development for the wider region and highlights the stakes for the individual riparians.

Firstly, this chapter briefly introduces the case study and its importance, before highlighting some of the important bilateral and multilateral historical developments, including the Argentinean–Brazilian relationship and its impact on regional institutions such as Mercosur. Following from this, the case study focuses on the treaties and agreements linked to the La Plata River Treaty as a key framework for future cooperation including all the five riparian member countries. The chapter will pay special attention to the traditional spheres of cooperation on the La Plata, mainly in the areas of navigation and hydrology, including key political, economic and strategic projects. This will include three key decisions made in recent times: the Itaipu Dam that was built by Paraguay and Brazil on the Parana River and is the largest producing hydropower dam in the world; the Hidrovia waterway transportation system reaching from Puerto Caceres in Brazil to Nueva Palmira in Uruguay covering over 3,400 kilometres; and the dispute between Uruguay and Argentina in regards to a pulp mill incident between 2005 and 2010.

Overall, there are similarities between the La Plata, Danube and Mekong rivers, including increased hydropower developments, importance of navigation (similar to the Danube) but also challenges of pollution and flooding. However, this chapter also showcases the uniqueness of the La Plata region in regards to several bilateral and trilateral treaties focusing on economic development and its impact on water resource management. Key challenges in the La Plata Basin include the less prominent role of public participation, underfinanced secretariats and a lack of communication and coordination between institutions. Similarly, the geographical positioning has a strong impact on cooperation and the institutions along the La Plata. The Comité Intergubernamental Coordinador de los Países de la Cuenca del Plata (CIC) is a rather weak forum and does not provide an adequately functioning umbrella. Similarly, Mercosur's role is limited and problems are dealt with on bilateral terms and are not included in the multilateral forums, even if all or most riparians are involved.

## Importance of the La Plata River Basin

The La Plata River Basin drains around one-fifth of the South American continent, and runs from north to south covering around 3.2 million square kilometres (see Map 5.1). In total, the La Plata River covers 17 per cent of the South American continent surface area, is the second largest river system in South America, after the better-known Amazon River, and is the fifth largest river in the world.

The river system comprises three sub-basins, namely the Paraguay, Parana and Uruguay River, and is the core of socio-economic activities in the region. The Paraguay River sub-basin hosts one of the highest numbers of endemic fishes, and the Parana River one of the highest number of endemic birds.[1] Further, the Pantanal wetlands, located in south-eastern Brazil, south-eastern Bolivia and northern Paraguay, are the largest wetlands in the world (including 650 species of birds, 260 species of fish, 90 species of reptiles, over 1,600 species of flowering plants and more than 80 species of mammals) and are currently being debated in domestic politics in Brazil and Bolivia.[2]

The La Plata River has an extremely variable topography ranging from 4,000 metres high in the mountains of Brazil to almost sea level in the southern plains of Argentina and Uruguay. Its tributaries arise in the Andes Mountain; rainfall varies from less than 700 mm per year in the western region to more than 1,800 mm in the Brazilian coastal ranges.[3] The Parana River is the most important river in the whole basin and supports three regional country economies, namely Brazil, Paraguay and Argentina. Combined, the Parana and Paraguay rivers account for around 84 per cent of the total La Plata River Basin and contribute around 80 per cent to the river water flow.[4] Overall, the basin includes 75 large dams, including the Parana River, which possesses one of the highest numbers of major dams worldwide.[5]

Hydropower development plays a key part in South America and has a high potential in all five riparian states (in a region with one of the greatest hydropower potentials globally). Current developments on the La Plata are driven by hydropower development (with potential currently exploited at around 60%) and urbanisation.[6] Internal migration increased the share of urban dwellers in

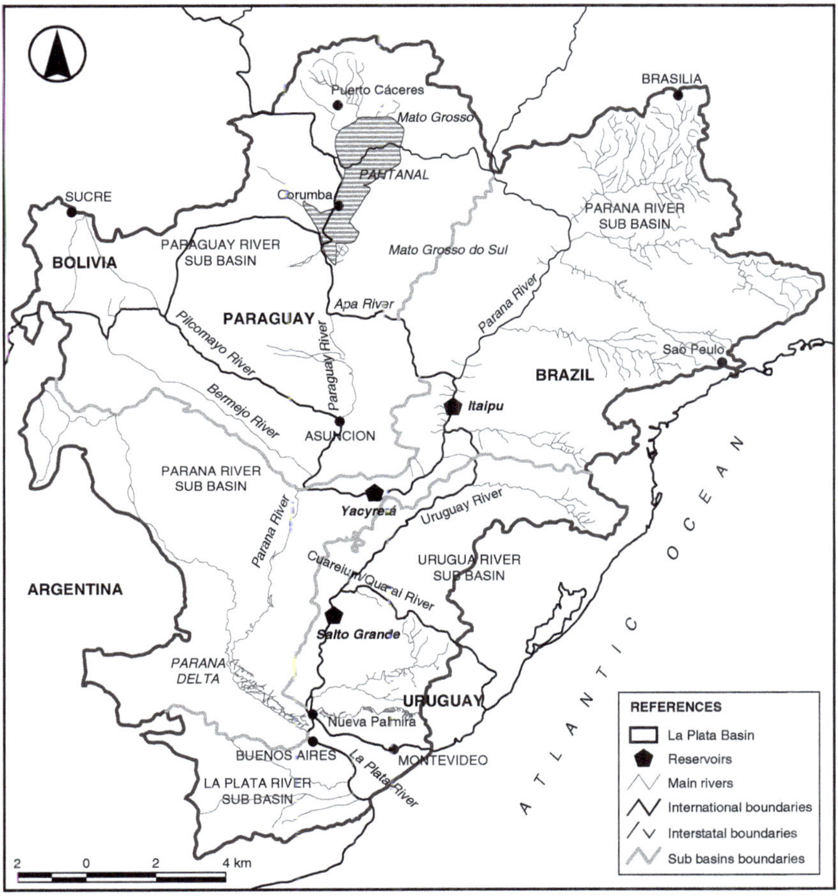


*Map 5.1* La Plata River Basin.

Source: Adjusted from Gilman *et al.*, 2008.

the region from 45 per cent in the 1960s to over 86 per cent in 2000s.[7] Due to the climatic conditions in the basin, rain-fed agriculture is widely used in the region and irrigated land is relatively low. However, agriculture – including soybean, maize and wheat – is the major water consumer and uses 62 per cent in Brazil and 96 per cent in Uruguay of the overall water consumption.[8]

Water quality and pollution levels on the La Plata are mainly affected by the increase in crops and industrial production, as well as population growth. Also, the changing land use from natural forest to soybean farming has affected the flow regime. Besides agriculture, fisheries and animal husbandry are key sources of food and income in the region. The basin is also rich in groundwater resources. The Guaraní aquifer, shared by all riparian countries except Bolivia, is one of the world's largest groundwater reservoirs. It extends over 1.19 million square kilometres and has an estimated capacity of 37,000 billion cubic metres, which are mainly used for consumption in over 300 Brazilian cities.[9]

Overall, the La Plata Basin is very important for economic development in South America, especially in the five riparian states, and also encompasses the most important industrialised areas of South America. On the other side, the La Plata is rated by the Word Wide Fund for Nature (WWF) as one of the top ten most threatened rivers in the world alongside the Mekong and Danube rivers.[10] The 27 additional major dams planned in the Brazilian part of the river combined with the Hidrovia navigation project would have a devastating impact on the environment in the region.

Combining historical and geopolitical events, according to the United Nations (UN) Water Report 2009, the key challenges in the La Plata River Basin include health issues, environmental challenges and risk management.[11] Insufficient sanitation and related infrastructure shortcomings are the main triggers for waterborne diseases such as diarrhoea, cholera, malaria or yellow fever. The environment is threatened through increases in population growth, mining, large water development projects and road development, which will result in the decrease of overall water quality. This also affects the erosion of land, silting of waterways and reservoirs, and soil and water pollution, causing loss of habitat for fish and wildlife, triggering biodiversity loss that is further intensified by the competing and often conflicting demands of the riparians on the limited regional natural resources.

## What Is at Stake for the Riparians

Nearly half of the La Plata River Basin falls within Brazil (45.9%), followed by Argentina (28.2%) and Paraguay (13.1%), with Uruguay and Bolivia sharing the remaining 12.8 per cent.[12] More than 50 large cities (with more than 100,000 people each), including the capital cities of Argentina, Brazil and Paraguay, are located along the La Plata Basin with a total population of over 100 million inhabitants.

Already, during the creation of the La Plata Treaty in 1969 and even more so today, the riparians in the La Plata River Basin have diverse interests and approaches towards the river and its management, which varies depending on the different historical, geographical, social and political backgrounds. The La Plata River Basin is a 'developing river basin and each country member has distinct water related demands and requirements'.[13] However, besides for the production of hydroelectricity and navigation, the La Plata River Basin is important to all the riparian countries for the use of water to support either the agricultural sector or the industrial sector. Rucks adds that 'historically and currently, water use in the Plata Basin has been driven by national goals, largely focused on generating economic benefits', a view that is still valid today.[14] Moreover, geopolitically, power relations play a crucial part in regards to cooperation and, similar to the Mekong and Danube river basins, the riparian countries in the La Plata vary significantly in socio-economic terms.

### *Brazil – The Regional Hegemon*

Brazil is considered one of the 'Amazon giants' and has the advantage of being the most upstream country in an area that has enormous natural resources. Brazil is the region's dominant political power and has been seen by Argentina in the past as a rival because of its size in relation to territory, population, economic growth and natural resources. In general, Brazil's interest in the La Plata is diverse and can be clustered into four areas. One key interest is the free navigation and transportation through the Hidrovia navigation canal, which supports especially the western regions of Brazil, including the Mato Grosso and Mato Grosso do Sul regions. In 2010 Brazil exported 15,140.268 tonnes on the Hidrovia navigation way of which nearly 50 per cent were soya exports, followed by mineral exports, corn, pulp and paper.[15] Imports through the Hidrovia are fewer, only 8,406.636 tonnes being imported by Brazil through the navigation channel; the largest percentage in this case (88%) was petroleum.[16] Brazil does not depend on the Hidrovia as much as Paraguay does, as it could create another linkage in its northern part. However, this would be very costly as Brazil would need to develop a new functional waterway and overall Hidrovia is the preferred and cheaper option as it is an already functioning waterway.

Another important aspect of the La Plata River for Brazil are the Pantanal wetlands – the world's largest floodplain – which are linked to the Paraguay River and are to 70 per cent within Brazil.[17] The Pantanal is a key hydrologic resource in South America and sustains flows in the Paraguay River throughout both the wet and dry season, which has a major impact on both the ecology and economics of the region between the Pantanal and the Atlantic Ocean. It is also an integral part of the hydrologic cycle of South America, due to its size and the quantity of water it temporarily stores annually. Increasing human impact, including dam construction, deforestation, agricultural activities and changes to the Hidrovia waterway, threaten the ecological stability of the Pantanal.

At the present time there are vigorous discussions about the future of the wetlands in Brazil. There are two options. The first option is the protection of the Pantanal wetlands, including its unique water flow that stores water from the Paraguay River for a long time, creating the wetlands and only releasing little water flow. The second option is the development of mining in the region (together with Bolivia), which will impact on the wetlands and the Paraguay River as it will change the water flow, thus affecting their size and function (as the Pantanal has a regulatory effect on the middle and lower Paraguay River discharge). While there is discussion within Brazil, at the international level and in relation to the other riparians Brazil is very conscious and supports fewer negative impacts in the Upper Basin.[18]

Hydropower is very important for Brazil as it is the major source of electricity; in 2010 it counted for around 74 per cent of the electrical energy consumed in Brazil, including electricity from bilateral projects such as the Itaipu Dam.[19] For instance, Brazil uses 90 per cent of electricity from the Itaipu Dam, which accounts for approximately 17 per cent of its energy usage and covers mainly the large cities, including Sao Paulo and Rio de Janeiro.[20] A fourth, broader interest of Brazil in regards to the La Plata is to show interest and cooperation with its neighbours to foster regional coherence and development through linked organisations such as the CIC, Mercosur or the Intergovernmental Committee for the Paraguay–Parana Waterway (CIH). As one expert has observed, 'Brazil is strong in the La Plata River not imposing but very diplomatic'.[21]

### *Bolivia – The Weak Upstream Country*

Bolivia is also an upstream country; however, it is not as powerful as Brazil. The country's relatively small size, its modest natural resources and landlocked position make it dependent on its neighbours on both sides of the La Plata River (although it is an upstream country in most of the smaller subbasins). However, in regards to the most important river in the western part, the Paraguay River, Bolivia is downstream of Brazil, which, combined with the weak institutional capacity of Bolivia and the lack of sufficient development, has decreased its negotiating power in the region. This has resulted in a lesser degree of involvement of Bolivia in the La Plata River even compared to Paraguay and Uruguay. Moreover, Bolivia's interest in the La Plata is rather low as the river only flows through the bottom part of the country which consists of mainly rural areas and attracts very little attention from the government, who mainly focuses on the Amazon River Basin and other rivers in the west of the country.

In general, Bolivia's interest in the La Plata River is to have access to the Atlantic. Bolivia does not have any interest in natural resource management as it is an upstream country, and also one of the major polluters on the Upper Basin due to mining activities in the border area with Brazil. Additionally, iron ore exportation and possible future plans could clash with Brazil's plans regarding the Pantanal wetlands as changes in the Upper Basin will impact on the unique floodplain system.[22] Another interest of Bolivia in the La Plata Basin is the Pilcomayo River, a tributary river, which has high levels of pollution. This is mainly the result of mining contamination that affects downstream countries, but also Bolivia itself, and is included in the Global Environment Facility (GEF) program as a pilot case study. Water pollution in this area, similar to the middle basin of the Danube River, can be traced back to the historical practice of mining.

### *Paraguay – Hydropower and Waterway of Key Importance*

Paraguay lies with all of its territory in the basin, which makes the basin very important for the country. Paraguay uses the Parana River for hydropower development based on mainly bilateral treaties. The Paraguay River is used for navigational purposes, mainly for agricultural products including soya and rice. As a result, Paraguay's power in the river basin rests not on political power, but on the reality that 50 per cent of the water necessary to power the hydroelectric dams is under its territory. Paraguay is still one of the poorest countries in South America and, similarly to Laos, depends significantly on the export of electricity to boost the economy and trade. Paraguay is the second largest power exporter and South America's only net exporter of hydroelectricity, which is mainly based on the Itaipu Dam system which also covers 75 per cent of Paraguay's own energy demand.[23] Besides the Itaipu Dam, which is run jointly by Brazil and Paraguay, Paraguay also has the Yacyreta hydropower dam which is shared with Argentina and a national dam, the Acaray Dam.

Hydropower production is important for Paraguay for two reasons. First, it covers the country's own energy demand (99%) with hydropower from the three dams and second (and more importantly), the export of energy to neighbouring countries and the revenues from it. This is also why negotiations in 2014 between Brazil and Paraguay regarding the rights for the Itaipu Dam were important for the country and a central focus of Paraguayan politics. Similarly, re-negotiations between Paraguay and Argentina regarding the Yacyreta Dam followed in 2015 and are a key focus of politics.

Besides hydropower, access to the sea via the Hidrovia channel is vital for Paraguay as it is a landlocked country. As a result, the Paraguay River is dominated by Paraguayan shipping companies along the Hidrovia, with estimates that around 80–90 per cent of the cargo ships are sailing under the Paraguayan flag. Additionally, almost 85 per cent of Paraguay's foreign trade is handled by the waterway.[24] Another challenge for Paraguay is the pollution originating from upstream countries Brazil and Bolivia. In general, Paraguay uses bilateral treaties more than multilateral ones, as it can be seen in the case of Itaipu and Yacyreta hydropower dams. These political negotiations are vital for the Paraguayan economy and are part of daily politics and election campaigns that further demonstrate the importance of the La Plata River Basin for Paraguay.

### *Uruguay – Downstream Country with Focus on Water Quality and Navigation*

Uruguay, together with Argentina, is a downstream country and one of the more prosperous states in South America. Uruguay's economy has been based on its strong agricultural sector, whose growth depends on the water from the La Plata. However, its power is relatively weak in the basin as a border country between Brazil and Argentina with close interpersonal and commercial ties to its neighbours. In general, Uruguay's interest in regards to the La Plata River is twofold: first, the access to the volume of water for hydropower generation; and, second, the water quality, which is especially important for downstream countries.[25]

Uruguay depends highly on hydropower for its electricity generation and accounts for a considerable share of energy production. For instance, the estimated use of hydropower in Uruguay in 2015 was around 51 per cent of the total energy consumption with the Salto Grande (a jointly managed dam with Argentina) on the Uruguay River providing a substantial amount besides three other large hydropower dams.[26] Further, water quality in Uruguay is slowly decreasing due to the release of untreated emissions from cities and industries besides agricultural runoff. Moreover, uncontrolled groundwater use can lead to pollution or overexploitation of aquifers. As a result, Uruguay is one of the main countries pushing for basin-wide monitoring of water quality and quantity. However, this is problematic as the standards and regulations in the five riparians vary depending on national policies.[27] This is one of the focal points of the GEF program, which is attempting to implement a common long-term monitoring program which should include fixed parameters.

Another current challenge exists between Uruguay and Argentina regarding a pulp mill incident on the Uruguay River lasting from 2005 until 2010. The case went in front of the International Court of Justice (ICJ) and still has implications today on the relations between these two countries and the bilateral body dealing with the Uruguay River, the Administrative Commission for the Uruguay River (CARU). Similarly, Uruguay and Argentina are competing for the continuous flow of cargo transportation charges from upstream countries that need to be reshipped for international transport in the ports of both countries. Argentina recently changed legislation and the modifications negatively impacted on Uruguay's ports; this shows that the challenges between both countries are still ongoing.[28]

### *Argentina – Downstream Country with Bargaining Power*

Argentina's interests are similar to the interests of Uruguay, as both riparians are the most downstream countries. These interests include a focus on water quality and enough water for hydropower generation. Water quality is important for Argentina to sustain tourism and agricultural activity. Additional concerns in Argentina include the impact through pollution and damage of the important fishing industry, especially in the area of the Uruguay River.

Uruguay and Argentina share the recent disputes on the Uruguay River regarding pollution, with Argentina recently changing the regulations for barges on the La Plata River. This has had a negative impact on Brazil and Paraguay as it restricts the size of barges and can result in economic loss as smaller size barges reduce the economic benefits. Argentina's reason for the decision is based on security concerns on the river; however, this could be a problem in the near future. Brazil's position is that regulations can be changed; however, it is important that the other countries are notified. Other recent changes by Argentina include the reshipping of cargo impacting on the use of Uruguayan ports that, as a result, are disadvantaged. In general, although Argentina has disadvantages regarding water pollution and other challenges resulting from being the most downstream country (along Uruguay), it can impact other countries in terms of the ability of navigation on the Hidrovia;

and, as an expert argued, the country uses this space to negotiate and bargain other issues in its economic interests.[29]

Overall, geographical power relations play a crucial part in regards to cooperation in transnational river basins and have a strong impact on the management of the water resources. However, in the La Plata Basin this is not as dominant as on the Danube River because downstream countries such as Argentina can influence the other countries by changing regulations on the Hidrovia waterway. Additionally, as highlighted in the next section, there is a strong commitment from all five countries to regional cooperation in recent times and that by working together all riparians can benefit from the La Plata River in different ways.

## Regional Developments in the La Plata Basin

Regional development in Latin America is built on a long history of conflict. The first half-century after independence was marked by post-colonial conflicts and wars of nation-building. This decreased in the late nineteenth century with only low numbers of interstate wars between South American countries. However, border disputes still existed, combined with long-running authoritarian regimes, enormous political instability and domestic violence.[30]

One of the key aspects in regards to regional development and basin-wide cooperation along the La Plata relates to the role of dominant regional players Brazil and Argentina, especially their relationship over time. An enormous change occurred in the bilateral relationship starting in the late 1970s. Brazil and Argentina's long-term rivalry was dominated by mutual hatred and military competition for regional leadership. For most of the past two centuries, Brazil and Argentina viewed each other suspiciously, competing for hegemony on the continent. However, this eventually evolved into a relationship of cooperative partnership, which was supported by the dynamics of democratisation and the economic crisis in both countries during the 1980s.[31] There have been several factors influencing this change, including the wish of both countries to reduce conflicts that in the past often resulted in military build-ups. Another factor is the common need to protect themselves from economic pressure. Argentina and Brazil realised during the late 1970s that due to the convergence in military and strategic interests, a peaceful resolution of outstanding disputes and the establishment of a durable framework for cooperation was mutually beneficial.

Key challenges during that time included an agreement about nuclear technology and the strategic value of the La Plata River for both countries. The La Plata River had strategic value for both countries, but the dispute was mainly about access and influence over the shared waterways, and not possession. In the 1970s, the rivalry increased due to the world energy crisis which made access and possession of natural resources a critical issue in the region, resulting in a resurgence of military competition and a stand-off between both rivals. The La Plata was a hot topic and the tension increased with the 1966 agreement between Brazil and Paraguay, which was not signed until 1973, to build the by-then world's largest hydropower dam, the Itaipu Dam system.[32] The key problem was the still ongoing struggle for influence over the neighbouring regional countries, highlighted by the signing of the Yacyreta Dam Treaty between Paraguay and Argentina only a few months later than the Itaipu Treaty. However, this challenge was solved by the tripartite Corpus–Itaipu Agreement between Brazil, Argentina and Paraguay which solved hydroelectrical disputes in the Upper Parana River, including the Itaipu and Yacyreta dams, and allowed both dams to run near the original planned capacity. However, scholars such as Resende-Santos argue that the agreement was more beneficial for Brazil and that Argentina surrendered its demands in the Upper Parana Basin.[33]

Overall, the 1979 dispute resolution over dam building rights on the Parana River can be seen as a turning point that opened the way for a flow of agreements affecting commerce, culture and nuclear energy. At the same time, Brazil and Argentina agreed to an economic integration plan that would replace imports of energy, food and machinery with local products. The economic package is also significant in broader geopolitical terms because cooperation between Brazil and Argentina provided the cornerstone for a broader regional common market involving Uruguay and Paraguay. Similarly, the Corpus–Itaipu Treaty was important as hydroelectric power and nuclear technology ended the military competition between Argentina and Brazil and established the first institutional structure. Overall, the agreement was used by both countries to concur on a 'flexible and agile mechanism' at the ministerial level to manage relations, which was important for future cooperation. These developments showed the close linkage of economic, security and political issues in the La Plata region and provided a framework for future bilateral and multilateral agreements.[34]

Further steps in the bilateral relationship included the signing of a nuclear safeguards agreement with the International Atomic Energy Agency (IAEA) in 1991, which helped to further constrain the nuclear ambitions of Argentina and Brazil and promoted bilateral confidence. By the end of the 1980s Brazil and Argentina had moved towards a common interest of defending democracy and international treatment for their foreign debt and trade. This 'reconciliation' provided a huge potential market for both countries in which the industrial, mining, energy and agricultural characteristics of both nations fostered complementarity instead of the limitations of the past driven by nationalistic rivalries and military ambitions.[35]

Overall, the democratisation process, economic integration and the Corpus–Itaipu Agreement played a key role in regional cooperative development. A key feature of cooperation between Brazil and Argentina is the linkage of security and economic issues besides the top-down process of decision-making (as most cooperation has been achieved through presidential meetings). Other than the merging of economic programs, the new, closer ties between Argentina and Brazil also included military, scientific, nuclear power as well as foreign policy issues. This is closely linked to regional developments as, noted by Brazil's deputy Foreign Minister in 1986, 'the path of Brazil–Argentina relations sets the rhythm for cooperation among all Latin American countries. If Argentina and Brazil are really able to proceed successfully, they will provide a good example to other Latin American countries'.[36] Similarly, more than 20 years later, in 2011, Argentina's Foreign Minister Hector Timerman said that both countries' relationship 'is indestructible' and that the good relationship is the key to 'bring stability to the region'.[37]

Besides these key developments between Argentina and Brazil, regional organisations in the 1960s included the Organisation of American States (OAS), which has been a principal vehicle after World War II and encompasses today all 35 independent countries of the Americas including the La Plata riparians. The OAS today constitutes the main political, juridical and social governmental forum in the hemisphere focusing on the main pillars of democracy, human rights, security and development.[38] In regards to developments on the La Plata River Basin, the OAS undertook a comprehensive study of its natural resources in the 1960s and 1970s. Today, the OAS is involved through its Office for Sustainable Development and Environment, the regional executive agency of the La Plata project, which is financed by the GEF and carried out in a collaboration between the CIC, the UN Environmental Program (UNEP), GEF (as the implementing agency) and the OAS.[39] Expanding on the OAS, regional development in South America has been integrated since the 1990s in a wider process of national and global economic development with Mercosur.

## *Southern Cone Common Market (Mercosur) – a Fading Regional Framework*

The 1988 Treaty of Integration, Cooperation and Development for the first time mentioned the common market as a key goal and included the establishment of a free trade agreement between both countries, Argentina and Brazil, within ten years. However, following this period, political changes and globalisation transformed the political environment and prompted an enlargement of the bilateral agreement into a larger regional association, including possible candidates Chile, Uruguay, Paraguay and Bolivia.[40] Negotiations between Argentina, Brazil, Paraguay and Uruguay on the common market continued and a key point was the institutionalisation of Mercosur. Whereas the smaller countries, Uruguay and Paraguay, supported supranational governance, Brazil as the biggest country did not support this and pushed for a common market.[41] In total, Brazil counts for approximately 70 per cent of the Gross Domestic Product (GDP) of Mercosur and as the regional hegemon has a major impact on the overall development and integration process.[42] Similarly, Argentina wanted to retain a certain degree of influence over the process and future developments. However, besides the uneven power distributions, negotiations and debates about the treaty text were held between all four parties, until all agreed and signed the Treaty of Asunción in 1991, establishing Mercosur.[43]

In 1994, Mercosur was a free trade area covering 95 per cent of intraregional trade.[44] In the same year, the Ouro Preto Protocol was added to the treaty and established a legal identity for the organisation as well as addressing institutional and dispute settlement issues. The protocol mandated in Article I the establishment of the institution with a Council of the Market, a Common Market Group, a Mercosur Trade Commission, Joint Parliamentary Commission, the Economic and Social Consultative Forum and the Mercosur Administrative Secretary.[45]

The Council of the Market inhabits the highest level, is responsible for the political management, the coordination of the common market, the integration progress and is also the decision-making organ (Section I, Article III).[46] The secretariat, which gives operational support to all the other organs, is relatively small and based in Montevideo, Uruguay (Section V, Article 31). It is funded by the member countries equally without an independent financial source.[47] However, although Mercosur's structure is similar to that of the Commission of the European Countries, its common bodies do not have enforcement tools. As a result, Mercosur cannot force member states to comply with the common market rules.

The decision-making process in Mercosur is based on the presidency of each country and all important decisions are made at the presidential summit. The Ouro Preto Protocol also established the customs union with a common external tariff covering 85 per cent of the goods traded with third countries, which in 1995 resulted in the second largest customs union in the world.[48] Further, Resolution No. 20/95 established ten subgroups of Mercosur including the Environment in Subgroup 6. That same year was the first time the environmental ministers of the countries met and accepted the Declaration of Taranco, highlighting the environment as a key sector in Mercosur agenda and institutions. However, it is important to stress that this declaration focuses on the inclusion of environmental matters in large development and infrastructure projects – similar to the Greater Mekong Subregion (GMS) in the Mekong region – and not on environmental issues and challenges in general.

### *Mercosur since 2000*

Mercosur has been seen by many as one of the most promising subregional integration projects of the 1990s. In 1998, the Mercosur member countries declared the area a zone of peace. However, the devaluation of the Brazilian currency in 1999 and 2001, together with the collapse of the Argentine economy in 2001, placed an enormous burden on Mercosur. This was further intensified by the appointment of a new Argentinian economic minister in 2001, who wanted to cooperate more closely with the United States (US) and pushed for a free trade area that would have allowed the individual countries to negotiate bilateral trade agreements.[49] This was combined with economic instability in Argentina that casted doubt on the future of Mercosur; the establishment of the common market was brought to a standstill.[50] All these problems were related to the structural problems of the organisation which existed from the beginning, such as the asymmetric power relations between Brazil and Argentina and the smaller neighbouring countries, and the problem of the exchange rates. Also, Argentina's view towards the US regarding bilateral ties played a role that contradicted the common market approach of Mercosur.

In mid-2000 Mercosur was relaunched, but it was still strongly affected by political uncertainties, the crisis in Argentina and economic problems in Brazil. Today Mercosur is dominated by Brazil, a 'natural leader of the region, in terms of both political stability and economic growth'; its steering of Mercosur is based on Brazilian politics and interests. Brazil uses Mercosur as a political and economic alliance to confront other powers outside the region such as the US or the European Union (EU) in the context of the Free Trade Area of the Americas (FTAA), the World Trade Organisation (WTO) or the EU–Mercosur debates.[51] As a result, Brazil's dominant position underpins the balance of power among the Mercosur countries. A major problem is the centralised institutional structure, as this is a disadvantage for the smaller member states that face political pressure by the larger states. Similarly, Mercosur is an institutionally weak organisation with insufficient political and administrative support.

Overall, the key interest of Brazil in regards to Mercosur remains increasing its international political weight through regional organisations. Cooperation with Argentina was considered an instrument to enhance political performance and economic competitiveness in a globalised world. Argentina's main interests were related to maximising economic growth, including integration with Brazil. Others argue that Mercosur was used by Brazil and Argentina to strengthen their bargaining power in regards to the North American Free Trade Agreement (NAFTA).[52]

Linking these findings to river basins, Gerald Berg (2011) argues that transnational regions such as the Southern Cone are important because economically integrated countries are more likely to cooperate in regards to border conflicts arising out of transnational river basins (such as the La Plata River Basin). Berg adds that the wish to mitigate border conflicts between Brazil and Argentina was one of the motivations behind Mercosur in the 1990s. Furthermore, there is potential for political alliances based on common interests as demonstrated by the economic integration provided by Mercosur, one that has reinforced the rapprochement between Brazil and Argentina, thus ensuring a more stable environment. In general, the common market reaffirms the dependency of the participant countries on Brazil. Also, Mercosur does not make any financial transfers to develop or diversify the other economies, as distinct from the EU system that supports less developed countries.

The less institutionalised option of Mercosur, without a supranational institution, makes it a very different case from the EU. Overall, the future of Mercosur depends on the convergence of interests and goals of both Brazil and Argentina. Possible future key challenges include a possible defection of Uruguay and a loss of interest from Brazil – the most important and dominant country – which recently has been pursuing key bilateral treaties, for instance, with the EU. A central part of Mercosur, similarly to the GMS in the Mekong region, is the increased integration of infrastructure projects and more intense transboundary ties between the riparian countries. Mercosur is also an example that shows how closely economics, politics and security are intertwined in South America.

Overall, Mercosur was a state-led approach – not driven by private sector or economic interdependence – and was seen by many as one of the most promising sub-regional integration projects of the 1990s. South America has been using formal regional trade treaties as the main channel of integration, in contrast to Southeast Asia which used a market approach. South American regional integration features a more mixed picture than Southeast Asian regional development. Mercosur, for instance, includes characteristics of 'old' regionalism, as seen in the state-led, mainly commercial and political initiative with little influence of civil society organisations.[53] However, it also includes features of the 'new' regionalism with its open and inclusive approach. South American regional integration is a 'politically led process with heavily institutionalized structures and forums, and antithetical position towards external markets' as related to the concept of open regionalism.[54] Open regionalism is also a common feature within Southeast Asian regional development. For instance, the Association of Southeast Asian Nations' (ASEAN's) economic integration supports close ties to China, South Korea and Japan. However, in regards to water management, compared to Asia's management of water bodies, South America is significantly less controversial and data sharing is less of a problem in comparison to other developing countries as in the case of the Mekong River.[55]

## Cooperation on the La Plata River Basin

Besides the regional frameworks, 15 treaties were signed between riparian countries over a 60-year period from 1946 to 2006, including three basin-wide multilateral agreements: the 1969 Treaty of La Plata Basin; the 1974 Treaty of the Financial Fund for the Development of the La Plata Basin (FONPLATA); and the 1991 Paraguay–Parana Waterway Treaty (which was followed by the 1992 Waterway Transport Agreement).

### *Treaty of La Plata Basin*

Negotiations on cooperation on the La Plata Basin started with the initiatives of Argentina and Brazil in the 1960s focusing on agricultural and industrial issues of the La Plata River Basin with the aim of supporting the necessary infrastructure. As a result of this initiative, in 1969 the five riparian countries signed the La Plata Basin Treaty. Article 1 of the treaty identifies the common areas of interest between the riparian states, including facilitation and assistance in navigation matters; the rational, fair and multiple use of water resources; and the promotion of projects of common interest, especially the assessment and good use of the regional natural resources.[56] The treaty includes a classical top-down diplomatic approach as negotiations are located at the ministerial level.

A major difference between the development on the La Plata – in comparison to the Mekong or Danube River – is the use of several bilateral and multilateral agreements between the riparian states besides the La Plata Treaty in order to achieve the general objectives of the basin development. For instance, Article IV of the La Plata Treaty states that commissions or secretariats can establish bilateral talks according to each riparian's laws and regulations. Further, collective action should not harm existing projects within the nations; each should follow international law and good practice of countries that are friends and neighbours.[57]

In general, the La Plata Treaty 'is not proposed as an exclusive option for riparian states, but as a framework agreement that could add special benefits to its global scheme'. Article VI of the treaty further states that the treaty will not inhibit any of the nations from conducting specific, partial, bilateral or multilateral agreements that pursue the general objectives of the river basin development.[58] This is one of the reasons why there exist several agreements in regards to water management in the basin.

### *Comité Intergubernamental Coordinador de los Países de la Cuenca del Plata (CIC)*

The creation of the CIC followed the La Plata Treaty in 1973, with a permanent secretariat in Buenos Aires. The purpose of the CIC is established in Article 3 of the La Plata Treaty, in which it is acknowledged as the

> permanent organ of the watershed, responsible for promoting, coordinating, and attending the functioning of multinational actions that aim at the integrated development of the La Plata Watershed, providing technical and financial assistance with support from international organizations considered advisable, and effect decisions which the Ministries of Foreign Affairs make.[59]

A key focus by then was on the reasonable utilisation of water resources and a basin-wide economic development. Similar to the approach of the GMS on the Mekong River, the treaty included projects beneficial for the region as a whole. The main goal was not only water management but the use of all natural resources in order to foster integration and economic growth between the riparian states.[60] During the 1990s other movements within the La Plata region affected and threatened the La Plata Treaty in its original form. These included mainly economic developments such as Mercosur, as the CIC and the early Mercosur had similar agendas that overlapped and competed. Although the CIC slowed down in its work during the early years of Mercosur, it persisted (along with the La Plata Treaty) and reaffirmed its importance, focusing more on the area of natural resources management and basin-wide development.

The CIC includes all five riparian states that have a share of the La Plata River Basin. During the La Plata River Treaty negotiations, Bolivia, Paraguay and Uruguay tried to transform the CIC into an international organisation with its own legal authority. However, Brazil and Argentina argued against this attempt and succeeded. Today, the CIC is seen as an advisory body which can recommend but cannot require actions from the member states, which provides a severe limitation similar to water management bodies of other river basins, including the Mekong River Commission (MRC). In general, in much the same way as the International Commission for the Protection of the Danube River (ICPDR) on the Danube River, the treaty provides an institutional base to discuss issues and raise concerns such as dam developments, transportation and exploitation of resources.

The CIC has two representatives from each country, one political and one technical, who meet twice a year. This is different to when the CIC was first created – the treaty was adjusted twice in 1992 and 2001, mainly to include the technical experts and to allow meetings of representatives outside the annual meetings to accelerate decision-making on administrative issues. Besides the ministers' annual meeting, extraordinary meetings can be held if necessary. Annual meetings at the Foreign Ministers' level started in 1969 when the institutional framework was established. In general, deliberation and decision-making is based on the unanimity rule which sometimes blocks the processes, as the affirmative vote of all five riparians is necessary in order to adopt a decision and no country can be absent. The Committee has, besides the administrative tasks, the responsibility to implement the Ministers' decisions and to enter into agreements with international organisations such as the World Bank, the United Nations Development Program (UNDP) or the GEF.

The CIC also includes a permanent secretariat based on Article 5 of the La Plata Treaty, responsible for the administration, communication and coordination of activities and projects between the member states (see Figure 5.1).[61] The General Secretary of the CIC is elected for a two-year term and can be re-elected for two additional terms of one year each. However, the candidature cannot exceed four years and a rotation occurs in alphabetical order of the riparian states.[62] Staff of the CIC is relatively stable and is based in Buenos Aires. The number of staff is small, similar to the number of staff at the ICPDR. There are also different secretaries-general for the major projects, including the current project on climate change and its impact on the La Plata, which is funded externally. In 2014, the total budget for the CIC included 261,700 United States Dollars (USD) and was financed internally by the riparian states. Brazil and Argentina paid 29.5 per cent, Uruguay 19 per cent, and Paraguay and Bolivia paid 11 per cent each.[63] There is additional funding for different programs and projects that are agreed upon separately and are organised differently through the riparian countries.

A positive aspect of the CIC is the high political status of the representatives. Overall, the La Plata agreements and treaties were successful in improving the economic development of all five riparians and in dealing with some transnational issues and challenges including hydropower and navigation. However, in

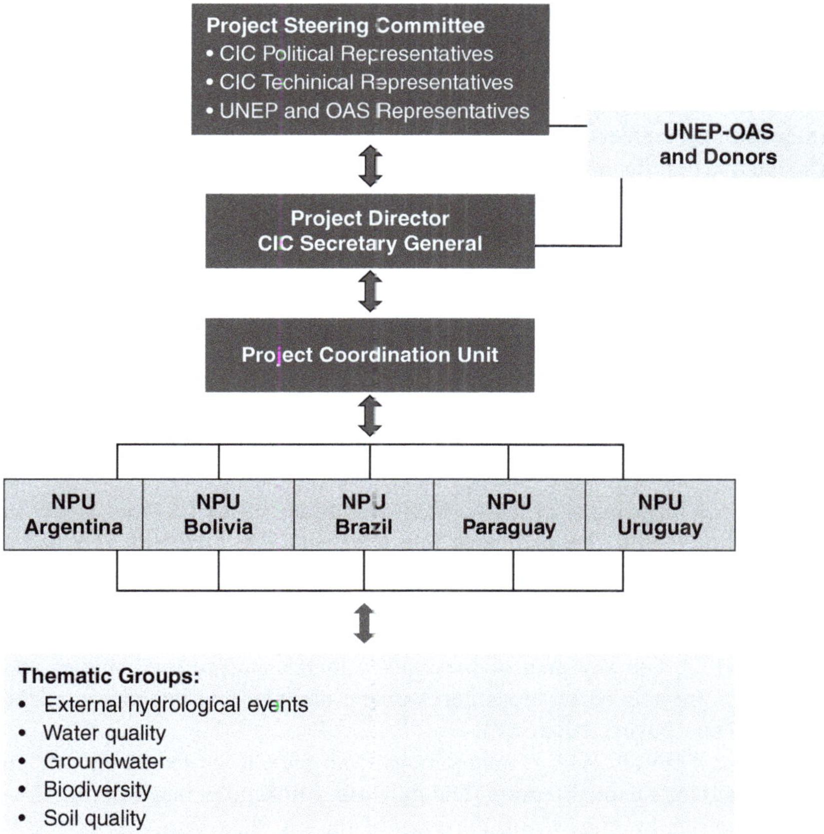


*Figure 5.1* CIC – La Plata Basin.

Source: adjusted from CIC, 2011.

NPU: National Planning Unit.

regards to other areas such as a basin-wide water management approach, the CIC has been unsuccessful. Moreover, for several decades the CIC lacked the inclusion of environmental issues such as deforestation or soil degradation.[64] Other key problems of the La Plata Basin include the understaffed Secretariat of the CIC and the lack of a functional technical department. Finally, the unanimity of quorums and decision-making to carry out any activity and an insufficient budget are additional challenges. Furthermore, more work is necessary to establish the CIC as a functional umbrella institution that coordinates activities and provides data for the whole basin. This also includes addressing the less-than-prominent role of public participation and the lack of communication between the different agreements and actions in order to achieve a basin-wide approach. Additionally, the treaty lacks the ability to make binding decisions on member states given that it lacks management and enforcement functions.

In comparison to the Mekong and Danube river basins, cooperation on the La Plata River has mainly focused on bilateral treaties and trilateral agreements instead of major regional or global agreements. This is also linked to South American developments in general as the region is known for its overlapping bilateral and multilateral cooperation in dealing with economic, diplomatic and security issues, which slows down efficiency. This includes a wide array of specific arrangements and parallel commissions that are different to other large basin developments and make it more challenging. Overall, in the La Plata region, 15 treaties were signed between riparian countries over a 60-year period, including only three basin-wide treaties that encompass all the riparians.

### *Financial Fund for the Development of the La Plata Basin (FONPLATA) and the Intergovernmental Committee for the Paraguay–Parana Waterway (CIH)*

The financial mechanism of the La Plata was added to the La Plata Treaty in 1974 as the FONPLATA and came into effect in 1976.[65] FONPLATA had a starting capital of 160 million USD, to which Argentina and Brazil contributed one third each, while the other one third is a combined contribution by Paraguay, Uruguay and Bolivia; the sum increased to 1,639.2 million USD by the end of 2013.[66] The key area of the fund is to finance projects, studies and research to promote harmonious development and physical integration of the La Plata River Basin (Article IV).

However, FONTPLATA is autonomous from the CIC in its administration and only submits annual reports. The only other linkage is that FONPLATA-funded projects must be along the lines of the La Plata Treaty objectives.[67] However, this demonstrates the lack of linkage between the different sections, institutions and commissions in the region. As a result, FONTPLATA does not finance the CIC secretary or its programs, which is controversial, especially in regards to the framework program financed by the GEF. One of the requirements for the financing was that FONTPLATA would contribute to the project; however, in the end FONTPLATA did not get involved. This raises further questions about the linkage of these regional institutions today.[68]

Besides FONTPLATA and the CIC, the Intergovernmental Committee for the CIH is the third multilateral agreement including all the five riparians of the La Plata River Basin. The agreement was signed in 1989 and established by the Ministers of Public Works and Transport, with its headquarters in Buenos Aires, Argentina.[69] The key focus of the CIH is navigation along the Hidrovia waterway, covering the Paraguay and Parana rivers which are around 3,400 kilometres long, reaching from Caceres in Brazil to Neuva Palmira in Uruguay.

The CIH has, similarly to the CIC, a Technical Commission and a Political Committee, with the former normally meeting three times a year and the latter twice.[70] However, there have not been any meetings since 2012 due to various reasons, including time management and unwillingness of the participants. The Chair of the CIH is rotated regularly and the CIH is assisted by an Executive Secretariat, which is temporarily financed by the Argentinean government. The Chief of Commission changes in alphabetical order and the Commission includes a representative from each country depending on the subsections that include, for instance, customs, dangerous goods, unifying of the nautical schools and others. The technical body discusses the technical problems and provides recommendations to the Political Committee that includes representatives from the Ministries of Foreign Affairs. The Political Committee makes the decision about the proposals made by the Technical Commission.

One of the key achievements is the Hidrovia Waterway Transport Agreement, which was incorporated into the La Plata Treaty, is a common navigation code for waterway users and is applicable to all five countries since its entry into force in 1995. As a result, all riparians have the same regulation to protect the water system and related ecosystems. However, a major challenge is the enforcement of these regulations as the implementation often varies significantly between the different countries. Some experts argue that the CIH is more important than the CIC as the focus is on transportation and is closely linked to economic gains, whereas the CIC focuses more on resource management with no immediate and direct monetary benefits.[71] Another challenge similar to the CIC is that decisions are made unanimously, which impacts on the effectiveness of the institution.[72] Likewise, due to bilateral issues or other challenges, there has not been any decision-making recently, which is a problem of the CIH and demonstrates its ineffectiveness.

### *Other Bilateral and Multilateral Treaties*

Besides the key treaties such as the La Plata River Basin and the Itaipu Treaty, the La Plata Basin comprises several other treaties coordinating actions and projects between two and more riparian countries. For instance, the 'Treaty on the La Plata River and its Maritime Front' was signed between Argentina and Uruguay in 1973, with a focus on settling the controversial situation between the countries over the river waters and also dealing with navigation, fishing, pollution and other scientific research. This resulted in two permanent commissions, namely the Administrative Commission for the La Plata and the Joint Technical Commission for the Maritime Front. Similarly, Argentina and Paraguay signed the Yacyreta Binational Entity in the same year, which dealt with another hydropower project, the Yacyreta development.

Also, in 1975 Argentina and Uruguay agreed on the establishment of a special body to deal with their common border along the Uruguay River that was followed by CARU. Brazil signed up with Argentina for a joint project regarding the Garabi Dam in 1980.[73] A similar wave of agreements followed between 1991 and 1995, with the most recent agreement being in 2006 between Brazil and Paraguay regarding the creation of the Brazil–Paraguay Commission of the Apa River Basin. Moreover, there are water-related agreements in all the major tributaries. The Parana River includes three partial commissions, namely the Parana River Joint Technical Commission between Argentina and Paraguay established in 1971; the Itaipu Bi-National Entity between Brazil and Paraguay in 1973; and the Yacyreta Bi-National Entity between Argentina and Paraguay in 1973. Further bilateral and trilateral agreements exist on the Rio de la Plata, the Pilcomayo River and the Bermejo River. Overall, all these agreements are autonomous and undertake different projects and programs. This makes a basin-wide approach very difficult and further demonstrates the distinctiveness of 'bilateral' regional cooperation.

In comparison to the Mekong and Danube River Basins, cooperation on the La Plata River has focused on bilateral treaties and agreements instead of multilateral ones. The agreements and treaties were, overall, successful in regards to the economic development of all the five riparians including hydropower and navigation. However, for several decades, they lacked the inclusion of environmental issues such as soil degradation. These issues are now being addressed by the new Strategic Action Plan that focuses on achieving a basin-wide integrated water resource management approach.

## The Itaipu Dam

After highlighting all the different agreements, treaties and organisations, this section focuses on the traditional spheres of cooperation on the La Plata, mainly in the areas of navigation and hydrology, including key political, economic and strategic projects linking the River Basin Organisation (RBO) level to specific case studies. The case studies include three key decisions made in recent time: the Itaipu Dam, the Hidrovia waterway and the pulp mill incident between Argentina and Uruguay. The first case study, the Itaipu Dam, was based on the Itaipu Treaty, which was signed in 1973 between Brazil and Paraguay and is today the second largest dam in the world (see Map 5.2). The treaty was signed during a time when Brazil was reaching for more regional power and Paraguay had suffered a defeat and became a landlocked state between the two regional powers, Argentina and Brazil.

During the 1960s Paraguay's only access to the world market was through the Paraguay and Parana rivers depending on Argentina. This was before Paraguay started cooperation on the Itaipu Dam system with Brazil that included international bridges to access the Atlantic coast in Brazil. Further, the building of the Itaipu Dam also solved the ongoing land and ownership dispute between both countries regarding the Guairá Falls, which were flooded and destroyed as a result of the dam.

Work on the Itaipu Dam started two years after the treaty was signed in 1975. The area of the dam is shared equally by Brazil and Paraguay and is classified as a 'special zone', which means that the territory is neutral, belonging to both countries. The Itaipu system covers nearly five miles across the Parana River with a height of 643 feet and a reservoir of 125 miles and 23.5 million acre-feet of water.[74] The Itaipu Dam took 20 years to be finalised as a project and seven

*Map 5.2* Itaipu Dam.

Source: Adjusted from Machain, 2009.

years to finish, with a total cost of more than 15 billion USD owned and shared equally by Brazil and Paraguay.[75]

The first turbine started running in 1984 and the whole dam was finished in 2007 running on full capacity with 20 turbines. At its earlier stage, 18 turbines were established but this was increased to 20, which have been fully running since 2007 reaching the maximum capacity of the dam. Ten turbines belong to each country, which is important as the frequencies are different between both countries: Brazil runs as the US on 60 hertz, whereas Paraguay uses 50 hertz like the rest of South America.[76] This also means that electricity sold from Itaipu, from Paraguay to Brazil, needs to be transformed before it can be used.

During the dam establishment, construction and housing in the border area increased significantly and around 80,000 workers (equally of both nationalities) built the dam; today 3,400 people are still employed by Itaipu Binational.[77] Currently, Paraguay uses around 10 per cent of the electricity from Itaipu that accounts for around 75 per cent of Paraguay's electricity. Brazil uses 90 per cent of the electricity, which counts for approximately 17 per cent of Brazil's energy usage and covers mainly the large cities Sao Paulo and Rio de Janeiro. Paraguay is the second largest power net exporter in the world, South America's only net exporter of hydroelectricity, and its main national agenda is closely tied to hydropower development as highlighted earlier.[78]

However, for a long time Paraguay was dissatisfied with the power exports from Itaipu to Brazil based on the Itaipu Treaty. The treaty restricted Paraguay's capacity to sell electricity to other markets; instead it stipulates that Brazil obtains its share of the electricity and the majority of Paraguay's share at a set price. According to Canese, 'Brazil obtains roughly 25 per cent of the electricity it consumes a year from Itaipu at one-third the international price'.[79] The key interests of Brazil (and also Argentina on the bilateral Yacyreta Dam) were based on the guarantee that a fixed price and quantity of power would be available in the future for an extended time (50 years on the Itaipu and 40 years on the Yacyreta). For the Itaipu Dam this means that in 2023, Paraguay could theoretically sell its share of energy to another country. However, this seems unlikely because to do so new transition lines would need to be built, which is expensive. Additionally, Brazil, although diversifying its energy matrix, still depends on the additional shares of the Itaipu Dam to support its growing energy demands.[80]

In 2004, Paraguayans started to demonstrate regularly outside the Brazilian embassy in the capital Asunción, demanding a re-negotiation of the treaty as Brazil payed a price to cover only the production cost.[81] Over time, the topic became a national issue in Paraguay before the dispute between both countries was solved with the signature of a new treaty in 2009.[82] This new agreement boosted Brazil's role as regional leader and showed that it is not an imperialist power towards smaller states as Brazil accepted the renegotiation and increased the payments towards Paraguay.

Additionally, it provided Paraguay with the needed increase in income from Itaipu and calmed down demonstrations and opposition against the initial treaty. Brazil tripled its annual payment to Paraguay for the power rights for the use of the dam from 124 to 360 million USD, adjusting the payment closer to the increasing market value of electrical power.[83] It should be noted that this does not include the kilowatt price that is still very low in comparison to worldwide prices.[84] However, the agreement allows Paraguay to sell the excess power directly to the Brazilian market at market rates as opposed to selling it only to Electrobras at below market rates; this is a significant victory for the Paraguayan government. After the new agreement, in 2009 the President of Paraguay Fernando Lugo declared on state television that 'for the first time in history, Brazil recognised our sovereignty and will contemplate the right for us to sell to other markets'.[85] For Paraguay this is an important step as it is the second poorest country in South America – although it has been exporting large amounts of power for more than three decades – and, similarly to Laos, it depends highly on the export of electricity to boost the economy and trade.

Besides the economic benefits and development of the massive dam project, negative effects of the Itaipu Dam included the drowning of the Guarina Falls, which were considered the greatest set of waterfalls in the world. Other negative impacts comprised the destruction of significant aquatic habitat, deforestation of 1,500 kilometres of jungle and forest on the Parana River and the displacement of around 42,000 people.[86] Another closely related challenge was the increase of water-borne diseases as a result of the reservoir's stagnant waters. Further, corruption was a major problem of the project with total cost estimated around 3.4 billion USD, costing eventually over 11 billion USD of which 20 per cent was sourced externally. As argued by Schilling and Canese, the Itaipu Dam construction was 'possibly the largest fraud in the history of capitalism'.[87]

Moreover, during the construction time of the dam until today, 149 people died in accidents, mainly during the construction time.[88] In contrast to the

Gabcikovo–Nagymaros Dam system on the Danube River, there was little opposition or resistance against the Itaipu Dam. This was mainly due to the focus on development during the time of the treaty and the high numbers of necessary labourers to build the dam. At the same time, environmental movements were less common and less organised, something that changed in the La Plata region later as will be seen in the case study of the pulp mills on the Uruguay River.

### *Regional Implications and Non-Involvement*

The Itaipu Dam system was planned in close proximity to Argentina's north on the Brazil–Paraguay border section of the Parana River. This triggered enormous protest from Argentina that resulted in a treaty signed with Paraguay only months later to build a second hydropower dam, the Yacyreta, close to the Itaipu Dam. The Yacyreta Dam was relatively small and less significant, but the key problem was that the height of the Itaipu plan would have negatively affected the Yacyreta Dam and its planned size. Itaipu plans would have also affected the crucial international trade and transport in the Argentinean part of the Parana River. Vice versa, Yacreya would have possibly flooded the turbines of Itaipu through increased water levels upstream.

These developments increased the heated debate about the La Plata River existing since the 1960s with only periodical but contentious negotiations. Argentina took the dispute further, resisting multilateral regulation and consultation, and involved the UN, the Non-Aligned Movement and regional forums such as Mercosur. Although some technical agreements between experts from both sides existed, the key problem was still the ongoing struggle for influence over the neighbouring regional countries. The first initiative to break down the stalemate came from Argentina who reversed its position at the end of 1976 and started talks with Brazil that resulted in the tripartite Corpus–Itaipu Agreement between Brazil, Argentina and Paraguay.[89] This resolved hydroelectrical disputes in the Upper Parana River, including Itaipu, Yacyreta and Corpus dam systems. The Itaipu Agreement was signed in Asunción, Paraguay, and allowed all three hydroelectrical systems to run near the original planned capacity. The purpose was to harmonise the Brazilian–Paraguay agreement on Itaipu with the Argentina–Paraguayan Yacyreta Treaty. The treaty sets the maximum operating level for the dams and its operational conditions.

### *Conclusion: Itaipu Dam*

To sum up, the Itaipu Treaty was signed during a time when Brazil was reaching for more regional power and Paraguay had suffered a major military defeat, becoming a landlocked state fully relying on Argentina for access to the world market. Additionally, the building of the Itaipu Dam solved the ongoing land and ownership dispute between Brazil and Paraguay regarding the Guairá Falls, which were flooded and destroyed as a result of the dam construction.

The Itaipu Dam was criticised, initially mostly by Argentina as the country feared possible consequences for the lower basin region and its hydropower plans. However, notwithstanding the criticism and the 1969 La Plata Agreement between the five riparians (over economic integration and joint development on the La Plata), altercations prevailed before the project went ahead. This affected the neighbourhood relationship – in a similar way to the Gabcikovo–Nagymaros Dam project on the Danube River.

Thereafter, the Itaipu Corpus Agreement provided the mechanism for more comprehensive water resource management among the three countries and reduced the protracted disputes along the Parana River. It further affected the integration and cooperation in other areas such as economic, social and political development in the Parana River hinterland. However, the example of the Itaipu Dam also shows that multilateral treaties play a limited or no role in contrast to bilateral treaties. Moreover, other regional bodies such as Mercosur and the CIC were not successful in solving the dispute between the riparians. In general, countries such as Paraguay prefer bilateral treaties in regards to commercial agreements, as this makes it easier to control and negotiate issues.

## Pulp Mill Incident on the Uruguay River

The second micro-level case study focuses on the pulp mill incident on the Uruguay River in the border area of Argentina and Uruguay. Both countries are neighbours with good historical ties of economic cooperation and political stability that changed in 2003, resulting in a more tense relationship.

Argentina and Uruguay became involved in an ongoing international dispute over the use of a common natural resource, lasting for several years with implications that still continue today. Already prior to 2005 there had been negotiations on building a pulp mill. However, in 2005 Uruguay authorised the construction of one of the largest pulp mills in the world in the border area with Argentina just five kilometres upstream from Fray Bentos, after negotiations were ongoing for a few years (see Map 5.3).[90] In 2005, the Finnish Company Botnia started construction on the site (the Orion (Botnia) mill) and the Spanish company Empresa Nacional de Celulose de Espana (CMB mill) received authorisation for a second plant mill that sparked severe opposition from the beginning.[91] Together, both mills accounted for an investment of 1.7 billion USD, with the International Financial Corporation (IFC) financing 175 million USD of the total and the Multilateral Investment Guarantee Agency guaranteeing up to 350 million USD.[92]

From the beginning of the project, it was attacked by Non-Governmental Organisations (NGOs) and other community groups for its lack of adherence to environmental and social standards. On both sides of the river, citizens worried about pollutants and toxins from the pulp mill affecting them and their environment. However, the difference was that in Uruguay many gained an income through jobs or exports from the mill, whereas in Argentina there were no benefits, just the fear of negative impacts on tourism and agriculture.

*Map 5.3* Uruguay River within the La Plata Basin.

Source: Adjusted from Petherick, 2010.

Additionally, concerns in Argentina included the impact from pollution and damage of the important fishing industry, an area of major concern for the government and the local population. Increased protest followed and in April 2005, 40,000 people gathered at the international bridge of 'Libertador San Martin' for a massive protest against the decision of the Uruguayan government. This resulted in the indefinite blockage of the international bridge between the countries, forcing commercial and tourist traffic to detour significantly.[93]

## *The Role of the Administrative Commission for the Uruguay River (CARU) and Other Institutions*

CARU[94] is the bilateral institution between Uruguay and Argentina dealing with the Uruguay River, established on the basis of the 1975 Statute of the Uruguay River (which became operative in 1976).[95] CARU was formally established in 1978 to handle the communications and work between both riparians with headquarters in Paysandú, Uruguay, and its own secretariat.[96] The above-mentioned pulp mill dispute between the countries mostly affected Article 7 of the Statute which states that a country that 'plans to construct new channels, substantially modify or alter existing ones or carry out any other works which are liable to affect navigation, the regime of the river or the quality of its waters, [it] shall notify the Commission'.[97] Further, if the Commission believes that significant damage could occur, the riparian needs to notify their counterpart and guarantee the opportunity to respond accordingly. This also includes the matter of pollution. Accordingly, Article 41 states that the parties must 'protect and preserve the aquatic environment and, in particular, to prevent its pollution, by prescribing appropriate rules and measures in accordance with applicable international agreements and in keeping, where relevant, with the guidelines and recommendations of international technical bodies'.[98]

Since its establishment and until the pulp mill incident, CARU never needed to have an active role in handling a conflict. However, during the concerns and escalation of the pulp mill incident, a High-Level Technical Group (GTAN) was created in 2005 to find a solution to the dispute. Yet, in early 2006 both riparians agreed that negotiations had failed and CARU was unable to prevent or resolve the conflict. As a result, Argentina submitted the dispute to the ICJ in May 2006 and also included a request for provisional measures, including preventing pollution and prescribing pollution control measures for the mill in accordance with international standards.[99] This was based on Articles 12 and 60 of the Statute, which states that any dispute which 'cannot be settled by direct negotiations may be submitted by either Party to the International Court of Justice'.[100] As a result, the case was submitted to the ICJ as a dispute mechanism over the interpretation and application of the existing treaty. The ineffectiveness of CARU was one of the key reasons for the escalation of the disagreement, besides the lack of representation, besides the lack of participation of local communities and non-governmental stakeholders.

In April 2010, the ICJ announced its judgement regarding the dispute between the two riparians and distinguished Argentina's claims as based on procedural claims and substantive claims. Whereas Argentina claimed that Uruguay breached all of its obligations under the terms of Articles 7 to 12 of the Statute, including both procedural and substantial claims, the ICJ ruled in regards to the procedural claims that Uruguay breached its obligation of notifying and consulting Argentina before authorising the pulp mill and construction.[101] However, the court also ruled that the declaration of Uruguay's breach was in itself a sufficient remedy for Argentina's claim and that Uruguay's action had not endangered the environmental health of the river, which means that the provisions were rejected by the court.

Another claim from Argentina was that Uruguay breached substantive treaty obligations to coordinate with Argentina through the bilateral river management agency (CARU) and to monitor and prevent pollution of the riverbed. However, the court found that no breach had occurred and rejected all other claims made.[102] Although Argentina won the procedural issues, the ICJ did not find any harm and, overall, Uruguay celebrated victory because it could promote pulp mills along the river. This resulted in economic gains for Uruguay and further investments as was seen, for instance, with the CMB mill that was restarted in a different location south of Fray Bentos after being abandoned in 2006. Similarly, the Botnia mill went ahead throughout the dispute and was fully functioning in 2007, converting wood chips into paper pulp since that time. Also, during the five years that it took to resolve the matter, trade and tourism between the two countries were severely damaged, impacting on the economies of both until the international bridge was reopened at the end of 2010 and both countries had a dramatic increase in tourists compared to the five years when the bridge was closed.

Besides CARU, Mercosur was also involved as Uruguay complained under the Mercosur framework that Argentina should take steps to remove protesters from the international bridge as this intervened with the rules of free flow of goods and people within Mercosur. As a result, Mercosur demanded that Argentina take steps to remove protesters. An *ad hoc* arbitration tribunal found that the blockades were not compatible with the Argentina's obligation under Mercosur, which included the guarantee of free circulation of goods and services.[103] However, similarly to CARU, Mercosur did not require Argentina to put an end to the blockade and failed to provide an effective remedy for the blockades.[104] Also similar to CARU was the fact that Mercosur was not able to successfully solve the problem of the blockade and did not put enough pressure on the Argentinean government to do so, which resulted in a claim from Uruguay to the ICJ based on provisional measures (as the blockade would interfere with the construction of the pulp mill).

Besides CARU and Mercosur, there has not been any involvement of the CIC during the dispute between both riparians, as the issue was dealt with in the bilateral arena of CARU or regionally through Mercosur. However, since 2014 the CIC is running a pilot project as part of the funded GEF program that looks into pollution measurement along the Uruguay River with the goal of achieving basin-wide standards in the future. Nevertheless, as previously highlighted in the Baia Mare incident on the Danube River, pollution measurements and standards are problematic to achieve as often the national measurements and levels vary from country to country resulting in different standards and legislations.

### *Conclusion: Pulp mill incident*

The pulp mill incident started in 2005 with Uruguay authorising one of the largest pulp mills in the world, which resulted in a decision by the ICJ in 2010 and has still implications today. The case highlighted the unsuccessful involvement of CARU, the bilateral institution dealing with the Uruguay River and the missed opportunity from Mercosur not succeeding with its interests regarding the re-opening of the international bridge linked to the pulp mill case. The case also showed that disputes over natural resources can have impacts on other areas of bilateral relationships. Although legitimising the pulp mill was beneficial for Uruguay and the international bridge is open again, the ICJ decision did not ease the tensions between both countries, similar to the ICJ decision on the Danube River regarding the Gabcikovo–Nagymaros Project.

## The Hidrovia Navigation Project

The last case study is the Hidrovia[105] project that was started under the La Plata treaty framework and later established the CIH which was briefly introduced previously. The Hidrovia waterway channel covers a 3,442 kilometre area along the Paraguay and Parana rivers, which renders the waterway navigable all year round from Puerto Caceres in Brazil to Nueva Palmira in Uruguay (see Map 5.4).[106]

*Map 5.4* Hidrovia navigation channel.

Source: Adjusted from CIC, 2014.

This allows cargo ships access to inland areas of Argentina, Brazil, Bolivia, Paraguay and Uruguay during the dry season that was previously not possible and further boosts the expansion of soybean, timber and iron ore trade.

The Hidrovia project was proposed in the late 1980s as a result of the increasingly necessary maintenance to the natural corridor and was the largest project for navigational river development worldwide.[107] The aim of the project was, and still is, similar to the Mekong River: to improve regional development by reducing transportation costs, improving links with commercial centres between large cities and providing landlocked nations (Bolivia and Paraguay) with access to the Atlantic Ocean. This also includes the rapidly increasing soybean production in the western parts of Brazil and the mining industry in Bolivia. The project would also directly link to the Tiete–Parana waterway in Brazil, which is 2,400 kilometres long and transports around two million tonnes of grains and other goods each year.[108]

The Paraguay–Parana waterway is naturally navigable based on its morphology. Unlike most of the great valleys of the world, navigation succeeds with a minimum slope that does not require the construction of systems of locks to allow transport on the river. The waterway has been sailed since historic times, first by aboriginal communities, and then by the conquerors, who used it to colonise a broad interior region of the continent.[109] Nowadays, navigation worldwide is increasingly affecting major rivers and associated wetlands due to channelling and modifying rivers through projects such as the Hidrovia impacting on, for instance, the Pantanal wetlands.

The Hidrovia project was planned since 1978 by regional foreign ministers and officially began in 1988 as a water superhighway. Around the same time, the CIH was established and linked to the CIC with the key focus of navigation along the Hidrovia waterway, covering the Paraguay and Parana rivers, with the goal of integrating the riparian countries. The CIH was established to promote and oversee the development of the commercial waterway and promote regional integration of the Mercosur countries. The Hidrovia Agreement links the riparian nations and strengthens regional integration among Mercosur members with the main objective of transforming the meandering river into a 2.5- to 3-metre-deep channel for the transit of barge convoys up to 60 metres in length, 24 hours a day, 365 days a year from Corumba in Brazil, Canal Tamengo in Bolivia and Santa Fe in Argentina.[110] The plan was to develop the navigational waterway along the 3,442 kilometres in two phases with the first phase including short-term interventions along 80 per cent of the project's length, and the second phase focusing on medium- and long-term interventions, including major river interventions along the entire channel.

After the project was proposed in the late 1980s, the Inter-American Development Bank (IDB) funded initial studies which were done and completed in 1990 by Internave, a Brazilian consulting firm, resulting in a positive assessment of the planned project.[111] This was followed by a study initiated by the CIH focusing on possible environmental impacts of the project (which were not meant to delay the project). However, in 1993 and 1994 environmental opposition started against

Hidrovia; it culminated in the creation of the Rios Vivos coalition,[112] a transnational coalition that included around 300 member organisations. Rios Vivos' main argument against the project was that it would destroy jobs and traditional ways of life, also reducing habitats and water quality.

As a result, between 1995 and 1997 new assessments were undertaken with the support of the IDB and the UNDP. These concluded that Hidrovia would benefit the region by reducing transportation costs, improving commerce and economic revenues, and stimulating industrial development, with little impact on the environment. These advantages were also the main reasons for the existence of other commercial waterways in the world, including the Danube River and the Nile River, amongst others. For instance, in the early 1990s the connection between the Rhine and Danube rivers in Europe was completed, linking the North Sea with the Black Sea.

In 1997, the Hidrovia project was still supported by the IDB as it was seen as a good opportunity for investment and regional development. Yet this changed after further reports and studies were done resulting in a termination of additional funds for the project by the IDB in the same year. The focus of the Hidrovia project shifted from a regional level to the national levels of the riparians, pushing it forward in Argentina, including investments of totalling 700 million USD in infrastructure by 2001 (with subsidies for dredging being cut in the same year due to the financial crisis in Argentina).[113] Simultaneously, Brazil had a negative standpoint on the whole project and focused on protecting the Pantanal, only doing small routine improvements and maintenance to keep the existing navigation consistent. Brazil has the least to gain from Hidrovia, and most of the new engineering work and new environmental impact would take place in Brazil. However, supporters argue that for Brazil it means an alternate and a more efficient export route for their increasing soy production in the Mato Grosso region.

In general, Hidrovia is important for all five riparian countries, but it is especially important for Bolivia and Paraguay, which will have access to the sea; also, Paraguay's placement of soybean production into foreign markets will be facilitated. Hidrovia also ensures that the complete fleets and ports are used and as the majority of barges – around 75 per cent in 2011 – sail under Paraguayan flag, this bring enormous economic benefits for the country.[114] For Argentina, as highlighted earlier, Hidrovia is important as 75 per cent of the population lives in the La Plata Basin, the area most affected by the project. For Bolivia, the project is the improvement of the output of their products to the La Plata, as an alternative to the land transport corridor to the Atlantic ports of Brazil, Chile and Peru in the Pacific. Uruguay benefits through the major ports for transhipment facilities and storage.

Opponents to the Hidrovia project have highlighted the enormous long-term costs that have resulted from similar projects, particularly in terms of 'externalities', including damages due to flooding, reduced drinking water quality, displaced populations, pollution, loss of wetlands and wildlife and damage to fisheries.[115] Others argue that already in the late 1990s the navigation traffic on the Paraguay River damaged levees and riparian vegetation, impacting on fishing communities. Additional devastating impacts on the environment in the whole river basin would include the modification and reduction of river flow, decrease of biodiversity, increase of water contamination and impact on the world's largest wetlands and the aquifer system.

These concerns were often raised by civil society or NGOs such as the Rios Vivos coalition who criticised the disseminated technical objections to the project, but also helped organise a broad-based coalition of environmental, social and indigenous organisations to discuss alternatives to the Hidrovia with local communities.[116] These concerns were further expanded through the increasing pollution and enhanced agriculture activity (including non-point sources), similar to the Danube Basin, which are major challenges for the river system. These contaminations are combined with industrial and heavy metal pollution from mining operations and other factories. Further, the proposed waterway would have a substantial impact on the flood regime of the Upper Paraguay River, affecting greater peak discharge in the river, reducing seasonal inundation of the floodplain and accelerating flood runoff. One of the major impacts would be that a small amount of decreased river levels would reduce floods, which would then have a strong effect on the floodplain ecosystem, especially during the low-water season when aquatic animals depend on a few remaining flooded areas for refuge.[117]

### *Role of the CIH and Mercosur*

A key problem of the CIH is the lack of a mechanism to include public participation in the decision process. This is similar to the CIC, which often lacks the public participation of civil society, NGOs or other interested groups; it also does not have a supra-legal authority. As a result, in the case of the Hidrovia project, the protest of more than 300 groups stopped the IDB investment in the project, as long as further environmental impact studies were done. The CIH also received criticism that its assessment study was incomplete and did not include long-term implications of water-level change on the floodplain, lacking both environmental assessment and public participation. Additionally, in 2013 Argentina changed the legislation which first reduced the size of cargo barges on the navigation channel due to safety reasons. Second, the legislation changed the number of transhipments from the river onto ocean vessels in the Uruguayan port of Palmeri. Both decisions have had consequences for the riparians, with Uruguay losing money through less transits, and Paraguay and Brazil being affected by the reduction of size for cargo barges on the channel.

In regards to Mercosur, Hidrovia has been very important as a key tool for regional industrial development, improving commerce for national and regional economies.[118] Additionally, the project links landlocked riparians with access to the sea and lowers the cost for transportation of goods, thus addressing three main areas of Mercosur: trade, transportation and navigation. Rivers are considered the axes of integration and cooperation processes,[119] which is why this case study is important in understanding Mercosur and other regional developments in South America. Overall, the waterway has become central to the process of internal integration of Mercosur focusing on the three aspects of navigation, trade and transport. Navigation on the La Plata has increased enormously and capacity, as well as traffic, is increasing yearly; shipping is possible all year around. Hidrovia increased the shipping from one million tonnes in 1992 to 16 million tonnes in 2011, with an ongoing increase in navigation.[120]

Positive results of the project included a fairly inexpensive way to push Mercosur's productivity and integration and the growth of cargo and operational capacity of transport equipment. Other benefits include a guarantee of permanent traffic; regularity that allows working with regular traffic based on the same schedule; competitiveness that promotes competition and encourages productivity; and security in terms of reduced interference, delays, congestion and accident risk.[121]

### *Conclusion: Hidrovia Waterway Project*

The Hidrovia waterway project can be seen as a starting point for regionalisation, covering economic, social, cultural, health and research areas, and, as a result, enhancing the growth of regional economies. Disadvantages of the Hidrovia include the negative effects on water, flora and fauna which are predicted, as well as the uprooting of the coastal population, the loss of potential for tourism, the spread of diseases, and, consequently, the deterioration of the quality of life of local people, including indigenous people. Other negative aspects include national bureaucratic indecision, problems with the free navigation of ships due to past different regulations by coastal guards, no unified protocols and lack of personnel training.

Overall, the CIH achieved the Waterway Transport Agreement, which was incorporated into the La Plata Treaty and is a common navigation code for waterway users and applicable to all five countries since its entry into force in 1995. As a result, all five riparians have regulation to protect the water system and related ecosystems. However, a major challenge is its enforcement and implementation, which often varies significantly between the different countries.

## Summary and Conclusion

South America has achieved rapid growth and yet this has resulted in major environmental challenges and problems. The La Plata River is significant for all the riparian countries and provides landlocked countries and regions with access to the Atlantic, as it can be seen with mining production in Bolivia or soybean production in the south-western parts of Brazil. Hydroelectrical use and navigation are still the key economic drivers of most bilateral and trilateral treaties and agreements, comprising projects such as Itaipu, Yacyreta and Hidrovia, and will further increase until all potential is exhausted.

The La Plata is similar to the Mekong and Danube River cases, including increased hydropower developments, the importance of navigation but also the challenges of pollution and flooding. However, the basin is also unique as it is dominated by several bilateral treaties and few multilateral treaties. Also, the role and framework of Mercosur is different to the EU or ASEAN, which together could be seen as disadvantages. However, a positive aspect of the La Plata is the low number of riparians – only five countries in comparison to 19 in the Danube River – and the participation of all of them in the CIC. This is not the case in the Mekong River, where the two upstream countries are not members of the MRC. Besides the differences, the La Plata has similarities with the other two case studies as the region was also confronted by wars, military dictatorships and colonial times in history. Similarly, there have been rivalries between the countries as highlighted by the transformed relationship between Brazil and Argentina from former enemies to partners, fostering a relationship that is widely regarded as promoting stability in the region.[122]

The first section highlighted the stakes of the La Plata riparians Bolivia, Brazil, Paraguay, Uruguay and Argentina and explained how the region and the relevant issues are interlinked. It also emphasised the dependence of some countries, especially Paraguay and Brazil, on the La Plata River for navigation and hydropower developments. Additionally, the first section demonstrated that asymmetric power relations in the different bilateral and multilateral agreements or even in Mercosur are important and show the national power interest of Brazil. However, in the La Plata Basin this is not as dominant as on the Danube River because downstream countries such as Argentina can influence the other countries by changing regulations on the Hidrovia waterway. Additionally, the case study showed that there is a strong commitment from all five countries to regional cooperation in recent times and that by working together all riparians can benefit from the La Plata River in different ways.

The chapter further showed that the CIC plays a rather inactive role when challenges occur; problems tend to be dealt with on bilateral terms rather than being dealt with in the multilateral forums (even if all or most riparians are included in these). This phenomenon is linked to the uniqueness of the La Plata River in comparison to the other two cases, as the basin includes several coexisting treaties. This is positive as it shows the institutionalisation of the region as a response to contested issues. However, it also complicates cooperation, especially due to the lack of an accepted and functional umbrella institution. Key problems include the limited role of public participation, underfinanced secretariats and a lack of communication between the different agreements and actions. All these are obstacles to the achievement of a basin-wide approach.

Further, the three micro-level case studies – the Itaipu Dam, the Hidrovia navigation channel and the pulp mill incident – show that in regards to economics and urgent issues, the bilateral way is the preferred option. It is clear that natural resource management and river cooperation are being used as political instruments; a demonstration of power can be seen, for instance, in the 'random' influence of Argentina on navigation and its consequential impact on the other riparians. Similarly, the pulp mill incident shows the role of Mercosur as a regional institution focusing on trade and the free flow of goods, which, however, could not resolve the dispute and lacked the enforcement to re-open the international bridge.

## Notes

1 Organization of American States (OAS), "La Plata River Basin," *Water Projects Series*, no. 6, 2005a.

2 Discussions are related to an increase in the length of navigation, which will result in a reduction of water retained in the wetland, possibly impacting on the ecosystem. See Jorge Rucks, "Experience and Role of the Organisation of American States in Transboundary River Basin Management in Latin America," in *Climate and Water – Transboundary Challenges in the Americas*, eds. Henry F. Diaz and Barbara J. Morehouse (Dordrecht: Kluwer Academic Publishers, 2003), 41–58; World Wide Fund for Nature (WWF), "World's Top 10 Rivers at Risk," WWF, Gland, 2007; Aaron T. Wolf and Joshua T. Newton, "Case Study Transboundary Dispute Resolution: The La Plata Basin," Institute for Water and Watersheds, n.d.-b.

3 Victor Pochat, "The La Plata River Basin," in *7th Biennial Rosenberg International Forum on Water Policy – Water for the Americas: Challenges and Opportunities* (Buenos Aires, November, 2010).

4 89.6 per cent of the drainage area is located in the Upper Parana Basin in Brazil. C. M. Krepper and V. Venturini, "Assessing Interannual Water Balance of La Plata River Basin," *Atmosfera* 22, no. 4 (2009): 389.

5 United Nations Educational, Scientific and Cultural Organisation (UNESCO), "United Nations World Water Development Report (WWDR 3): Water in a Changing World," UNESCO, Paris and London, 2009; OAS, "La Plata River Basin."

6 Victor Pochat, "International Agreements, Institutions and Projects in La Plata River Basin," *International Journal of Water Resources Development* 27, no. 3 (2011): 501.

7 Ibid.; UNESCO, "United Nations World Water Development Report (WWDR 3): Water in a Changing World."

8 Pochat, "International Agreements, Institutions and Projects in La Plata River Basin."

9 UNESCO, "United Nations World Water Development Report (WWDR 3): Water in a Changing World."

10 WWF, "World's Top 10 Rivers at Risk."

11 UNESCO, "United Nations World Water Development Report (WWDR 3): Water in a Changing World."

12 Arun P. Elhance, *Hydropolitics in the 3rd World – Cooperation and Conflict in International River Basins* (Washington, DC: United States Institute for Peace, 1999).

13 Rolando Quiros, "The Plata River Basin: International Basin Development and Riverine Fisheries," in *Proceedings of the Second International Symposium on the Management of Large Rivers for Fisheries Volume I*, eds. Ian G. Cowx, O. Almeida, C. Bene, R. Brummett, S. Bush, W. Darwall, J. Pittock, and M. van Brakel (Bangkok: FAO Regional Office for Asia and the Pacific, 2004), 266.

14 Rucks, "Experience and Role of the Organisation of American States in Transboundary River Basin Management in Latin America," 46.

15 Agência Nacional de Transportes Aquaviários (ANTAQ), "Bacia Do Paraguai – Plano Nacional De Integração Hidroviária [The Paraguay Basin – National Plan of Waterway Integration]," ANTAQ, Brasilia, 2013, 18.

16 Ibid.

17 Ana Lucia Lima Barros Dolabella, "The Brazilian Pantanal: An Overview," in *The Pantanal of Brazil, Bolivia and Paraguay*, ed. Frederick A. Swarts (Gouldsboro, PA: Hudson MacArthur Publishers, 2000), Chapter 3.

18 Interview with a senior expert of natural resources management in South America in Montevideo, Uruguay on the 28.04.2014.

19 Manlio Coviello, Juan Gollan, and Miguel Perez, "Las Alianzas Publico–Privados En Energias Renovables En America Latina Y Caribe [The Private–Public Renewable Energy Alliance in Latin America and the Caribbean]," CEPAL, Santiago de Chile, 2012, 11.
20 Interview with an Itaipu expert in Foz do Iguacu on the 04.05.2014.
21 Interview with a senior expert of the CIC in Buenos Aires on the 24.04.2014.
22 Interview with a senior expert of natural resources management in South America in Montevideo, Uruguay on the 28.04.2014.
23 Jane Monahan, "Paraguay Looks to Stem Flow of Cheap Power from Itaipu Dam to Brazil," *Energy Economist*, no. 286 (2005): 14–16.
24 Guillermo Haskel, "Paraguay Calls on Neighbours to Revamp Hidrovia Fleet and Ports," *Buenos Aires Herald*, 12.05.2014.
25 Interview with a senior expert of natural resources management in South America in Montevideo, Uruguay on the 28.04.2014.
26 Centro de Formación en Integracion Regional (CEFIR), "Atlas de Energias Renovables del Mercosur [Atlas of Renewable Energies in Mercosur]," www.cefir.org.uy/ (accessed 03.06.2014).
27 Skype interview with an expert of the CIC on the 29.05.2014.
28 Interview with a senior expert on navigation and transportation in Montevideo, Uruguay on the 28.04.2014.
29 Interview with a senior expert on navigation and transportation in Montevideo, Uruguay on the 28.04.2014.
30 CEFIR, "Atlas de Energias Renovables del Mercosur [Atlas of Renewable Energies in Mercosur]," 531.
31 Joao Resende-Santos, "The Origins of Security Cooperation in the Southern Cone," *Latin American Politics and Society* 44, no. 4 (2002): 92.
32 During this time the Itaipu Dam system was the world's largest system; however, today it is the second largest after the Three Gorges Dam system in China. Resende-Santos, "The Origins of Security Cooperation in the Southern Cone."
33 Resende-Santos, "The Origins of Security Cooperation in the Southern Cone."
34 Ibid.
35 Juan De Onis, "Brazil, Argentina Agree on Economic Integration Plan," *Los Angeles Times*, 14.07.1986.
36 Ibid., 1.
37 Hector Timerman, "Our Alliance with Brazil Is Indestructible," *Buenos Aires Herald*, 30.01.2011.
38 OAS, "OAS – Who We Are," www.oas.org/en/about/who_we_are.asp (accessed 20.05.2014).
39 OAS, "La Plata River Basin – A Framework for the Sustainable Management of its Water Resources with Respect to the Hydrological Effects of Climate Variability and Change," *Water Projects Series*, 2005b.
40 Chile was not interested in such an association as it worried about the instability of its neighbouring countries and Bolivia was not accepted because it was a member of the Andean Pact and predisposed towards another common market. Uruguay had already been an observer to meetings of the integration process in the past and its economy was closely linked to the one of its big neighbours. Similarly, Paraguay joined the association after the return to democracy and negotiations in 1990. Paraguay was an asset to Brazil as it is an important provider of cheap labour which is indispensable for the southern states, whilst Argentina was interested in electricity supply from the neighbour.
41 Brazil is the largest country with around 75 per cent of total assets including GDP, population and trade. In comparison, Germany, as the largest member in the EU, only holds around 33 per cent of all assets. Andy Klom, "Mercosur and Brazil: A European Perspective," *International Affairs* 79, no. 2 (2003): 352–3.

42 Pedro Motta Veiga, "Brazil in Mercosur: Reciprocal Influence," in *Mercosur: Regional Integration, World Markets*, ed. Riordan Roett (Boulder, CO: Lynne Rienner, 1999), 25–33.
43 Venezuela joined the grouping in 2006 as a full member; however, this was not ratified until 2012. This has been closely linked to the ban of Paraguay's officials to Mercosur due to a political coup in 2012. Paraguay was rejecting Venezuela's full integration into Mercosur since 2006 and blocked it in all referendums in the Paraguayan Congress since Venezuela's membership status in 2006. However, due to the dismissal of the Paraguayan president in the senate in an impeachment trial and the following ban of the Paraguayan officials in Mercosur, Paraguay could not oppose the full integration of Venezuela and the country joined Mercosur in mid-2012. Mercosur, "Quienes Somos [Who We Are]," www.mercosur.int/innovaportal/v/6283/2/innova.front/quienes-somos (accessed 09.10.2012); Ruxandra Adam, "Venezuela Is Officially Accepted in Mercosur," *Softpedia*, 22.07.2006; Guido Nejamkis and Ana Flor, "Mercosur Welcomes Venezuela, Suspends Paraguay," *Reuters.com*, 29.07.2012.
44 Klom, "Mercosur and Brazil: A European Perspective."
45 Mercosur, "Protocolo De Ouro Preto [Ouro Preto Protocol]," www.mercosur.int/innovaportal/file/721/1/cmc_1994_protocolo_ouro_preto_es.pdf (accessed 09.10.2012).
46 Ibid., Section I, Article III.
47 Ibid., Section V, Article XXXI; Sam Laird, "Mercosur: Objectives and Achievements," World Trade Organization (Trade Policy Review Division), Geneva, 1997.
48 Ibid.
49 Klom, "Mercosur and Brazil: A European Perspective"; Mario E. Carranza, "Can Mercosur Survive? Domestic and International Constraints on Mercosur," *Latin American Politics and Society* 45, no. 2 (2003): 85.
50 Klom, "Mercosur and Brazil: A European Perspective."
51 Ibid., 355.
52 Carranza, "Can Mercosur Survive? Domestic and International Constraints on Mercosur."
53 Laura Gomez-Mera, "How 'New' Is the 'New Regionalism' in the Americas? The Case of Mercosur," *Journal of International Relations and Development* 11, no. 3 (2008): 282.
54 Mauricio Baquero-Herrera, "Latin America Update: Something Is Going on in Latin America," *Law and Business of the Americas* 12, no. 4 (2006): 615; Gomez-Mera, "How 'New' Is the 'New Regionalism' in the Americas? The Case of Mercosur."
55 Asti K. Biswas, "Transboundary Water Management in Latin America: Personal Reflections," *International Journal of Water Resources Development* 27, no. 3 (2011b): 423–9.
56 Comité Intergubernamental Coordinador de los Países de la Cuenca del Plata (CIC), "El Tratado De La Cuenca Del Plata [Treaty of the La Plata Basin]," CIC, Brasilia, 1969, Article I.
57 CIC, "El Tratado De La Cuenca Del Plata [Treaty of the La Plata Basin]," Articles IV and V.
58 CIC, "El Tratado De La Cuenca Del Plata [Treaty of the La Plata Basin]," Article VI.
59 CIC, "El Tratado De La Cuenca Del Plata [Treaty of the La Plata Basin]," Article III.
60 Lilian Del Castillo, "The La Plata Basin System against the Background of Other Basin Organizations," *International Journal of Water Resources Development* 27, no. 3 (2011): 515.
61 Ibid.; CIC, "El Tratado De La Cuenca Del Plata [Treaty of the La Plata Basin]."
62 Interview with a senior expert on the politics of the La Plata River in Buenos Aires on the 24.04.2014.

63 Interview with a political representative to the CIC in Buenos Aires on the 22.04.2014.
64 Patrick Gilman, Victor Pochat, and Ariel Dinar, "Whiter La Plata? Assessing the State of Transboundary Water Resource Cooperation in the Basin," *Natural Resources Forum* 32, no. 3 (2008): 205.
65 Funded projects include the Framework Program for Water Assessment and Climatic Variability or the Parana–Paraguay Waterway.
66 Treaty of the Financial Fund for the Development of the La Plata Basin (FONPLATA), "Convenio Constitutivo Del Fondo Financiero Para El Desarrollo De La Cuenca Del Plata [Agreement of the Financial Fund for the Development of the La Plata Basin]," FONPLATA, Buenos Aires, 1974, Article IV, 3; FONPLATA, "Estructura del Capital [Structure of the Capital]," www.fonplata.org/index.php/fonplata/nuestra-historia (accessed 21.04.2015).
67 Del Castillo, "The La Plata Basin System against the Background of Other Basin Organizations"; FONPLATA, "Memoria Institucional Fonplata 2010 [2010 Annual Report of the Fonplata Institution]," FONPLATA, Santa Cruz de la Sierra, Bolivia, 2011.
68 Skype interview with a senior CIC staff on the 29.05.2014.
69 Victor M. Ponce, "Geographical Background, Pantanal and Hidrovia," in *Hidrovia Report – Wetland Hydrology*, San Diego State University, San Diego, CA, n.d.
70 Interview with a CIH representative in Montevideo, Uruguay on the 28.04.2014.
71 Senior Expert on water resources management in South America, Montevideo, Uruguay on the 28.04.2014.
72 CIC, "Programa Hidrovia Paraguay–Parana [Paraguay–Parana Waterway Program]," www.cicplata.org/?id=hpp_intro (accessed 19.06.2014).
73 For a full list of all bilateral, trilateral and multilateral agreements see the tables in Pochat (2011: 504).
74 Monahan, "Paraguay Looks to Stem Flow of Cheap Power from Itaipu Dam to Brazil."
75 Elhance, *Hydropolitics in the 3rd World – Cooperation and Conflict in International River Basins*.
76 Interview with an Itaipu expert in Foz do Iguacu on the 04.05.2014.
77 The Itaipu Binational Company was established based on the Itaipu Treaty and is half owned by Paraguay's state-owned electric utility Administración Nacional de Electricidad (ANDE) and half by Brazil's federal holding company Electrobras.
78 Santa B. Pun, "Paraguay, Bhutan and Nepal: Landlocked but Hydropower Rich," *Hydro Nepal* 3, no. 3 (2008): 1–5.
79 Quoted in Monahan, "Paraguay Looks to Stem Flow of Cheap Power from Itaipu Dam to Brazil," 14.
80 Interview with an Itaipu expert in Foz do Iguacu on the 04.05.2014.
81 Monahan, "Paraguay Looks to Stem Flow of Cheap Power from Itaipu Dam to Brazil"; Raul Zibechi, "Is Brazil Creating its Own 'Backyard'?," Center for International Policy, Washington, DC, 2009, 2.
82 J. Eliseo Da Rosa, "Economics, Politics, and Hydroelectric Power: The Parana River Basin," *Latin American Research Review* 18, no. 3 (1983): 80.
83 The Economist Intelligence Unit, "Brazil/Paraguay Politics: A Dam Deal," *EIU Views Wire*, 27.07.2009; Da Rosa, "Economics, Politics, and Hydroelectric Power: The Parana River Basin."
84 Interview with an Itaipu expert in Foz do Iguacu on the 04.05.2014.
85 Pun, "Paraguay, Bhutan and Nepal: Landlocked but Hydropower Rich," 3.
86 WWF, "World's Top 10 Rivers at Risk"; Monahan, "Paraguay Looks to Stem Flow of Cheap Power from Itaipu Dam to Brazil."
87 Quoted in Monahan, "Paraguay Looks to Stem Flow of Cheap Power from Itaipu Dam to Brazil," 14.

88 Interview with an Itaipu expert in Foz do Iguacu on the 04.05.2014.
89 According to Resende-Santos, the three key reasons for the change in position include the crisis in the South Atlantic, a growing power disparity between Argentina and Brazil and the renewed Argentinean concerns about the US–Brazil relationship. Resende-Santos, "The Origins of Security Cooperation in the Southern Cone," 99.
90 Anna Petherick, "Science Panel Gives Hope in River-Pollution Dispute," *Nature* 466, no. 911 (2010): 10.
91 Donald K. Anton, "Case Concerning Pulp Mills on the River Uruguay," *Australian International Law Journal* 17 (2010): 213–23; Jacob K. Cogan, "The 2010 Judicial Activity of the International Court of Justice," *The American Journal of International Law* 105, no. 3 (2011): 479; Patricia Avila, "Uruguay Approves Metsa–Botnia Pulp Mill," www.reuters.com/article/2007/11/10/environment-metsabotnia-uruguay-dc-idUSL0941462920071110 (accessed 05.05.2014).
92 The IFC is a section of the World Bank Group financing private sector investment, facilitating access to capital and advising business and government. Thilo Rensmann, "Recent Development in the Jurisprudence of the International Court of Justice," TU Dresden, Dresden, 2010.
93 Ramiro Berardo and Andrea K. Gerlak, "Conflict and Cooperation Along International Rivers: Crafting a Model of Institutional Effectiveness," *Global Environmental Politics* 12, no. 1 (2012): 105.
94 In Spanish: Comisión Administradora del Rio Uruguay.
95 Administrative Commission for the Uruguay River (CARU), "The River Uruguay Executive Commission," CARU, Paysandu, Uruguay, n.d.
96 CARU, "Statute of the River Uruguay. Signed at Salto on 26 February 1975," *United Nations – Treaty Series*, 1975, Articles 49–52.
97 Ibid., Article 7.
98 Ibid., Article 41a.
99 Rensmann, "Recent Development in the Jurisprudence of the International Court of Justice."
100 CARU, "Statute of the River Uruguay. Signed at Salto on 26 February 1975," Articles 12, 60; CARU, "The River Uruguay Executive Commission."
101 International Court of Justice (ICJ), "Press Release: Pulp Mills on the River Uruguay (Argentina V. Uruguay)," ICJ, The Hague, 2010b.
102 ICJ, "Press Release: Pulp Mills on the River Uruguay (Argentina V. Uruguay)," 1.
103 Rensmann, "Recent Development in the Jurisprudence of the International Court of Justice."
104 Cymie R. Payne, "Environmental Impact Assessment as Duty under International Law: The International Court of Justice Judgement on Pulp Mills on the River Uruguay," *The European Journal of Risk Regulation* 3 (2010): 317–24.
105 'Hidrovia' means waterway in Spanish and Portuguese.
106 Gilman, Pochat, and Dinar, "Whiter La Plata? Assessing the State of Transboundary Water Resource Cooperation in the Basin."
107 UNESCO, "United Nations World Water Development Report (WWDR 3): Water in a Changing World"; Wolf and Newton, "Case Study Transboundary Dispute Resolution: The La Plata Basin."
108 Pochat, "International Agreements, Institutions and Projects in La Plata River Basin."
109 CIC, "Programa Hidrovia Paraguay–Parana [Waterway Paraguay–Parana Program]"; Griselda Capaldo, "The Regional Integration Process," *Environmental Policy and Law* 43, nos. 4–5 (2013): 222–5.
110 Capaldo, "The Regional Integration Process"; Johan F. Gottgens, James E. Perry, Ronald H. Fortney, Jill E. Meyer, Michael Benedict, and Brian E. Rood, "The Paraguay–Parana Hidrovia: Protecting the Pantanal with Lessons from the Past," *BioScience* 51, no. 4 (2001): 304.

111 Ibid.
112 Translated in English: 'Living Rivers'.
113 Ibid.
114 Alvaro Galli, "La Experencia De La Hidrovia Paraguay–Parana [The Experience of the Paraguay–Parana Waterway]," in *Navegacion y Control de Inundaciones en el Rio Magdalena: Un Proyeto Pais [Navigation and Control of the Flooding in the Magdalena River: A Country Project]* (Bogota, Colombia, September, 2011); Maria Eugenia Martinez, Martina Bonadeo, and Matias Barros, "Hidrovia Parana–Paraguay; Mercosur y el Comercio [the Paraguay–Parana Waterway; Mercosur and Trading]," www.slideshare.net/Jenny1994/hidrovia-paranparaguay-mara-eugenia-martnez-martina-bonadeo-y-matas-barrios (accessed 10.08.2014).
115 Johan F. Gottgens, "The Paraguay–Parana Hidrovia: Large Scale Channelization or a 'Tyranny of Small Decision'," in *The Pantanal of Brazil, Bolivia and Paraguay*, ed. Frederick A. Swarts (Gouldsboro, PA: Hudson MacArthur Publishers, 2000), Chapter 14.
116 International Rivers Network, "Paraguay–Parana Hidrovia," www.internationalrivers.org/campaigns/paraguay-paran%C3%A1-hidrovia (accessed 10.08.2014).
117 Stephen K. Hamilton, "Potential Effects of a Major Navigation Project (Paraguay–Parana Hidrovia) on Inundation in the Pantanal Floodplains," *Regulated Rivers: Research & Management* 15, no. 4 (1999): 296.
118 Martinez, Bonadeo, and Barros, "Hidrovia Parana–Paraguay; Mercosur y el Comercio [the Paraguay–Parana Waterway; Mercosur and Trading]," 7.
119 Capaldo, "The Regional Integration Process."
120 Galli, "La Experencia De La Hidrovia Paraguay–Parana [the Experience of the Paraguay–Parana Waterway]."
121 Damaris Barreto, Pamela Furrer, Auro Guerra, Monica Hernandez, Angelica Mateo, and Rodolfo Martinez, "Hidrovias Del Mercosur: De La Ilusion Al Desencanto [Waterways of Mercosur: From Illusion to Disenchantment]," National University of Litoral, Santa Fe, Argentina, 2008.
122 Timerman, "Our Alliance with Brazil Is Indestructible."

# 6 Cross-Case Analysis, Findings and Lessons

This book has investigated the phenomenon of cooperation and conflict over transboundary water resources. In particular, it has focused on transboundary river cooperation between riparian states with asymmetrical power relations, where the dominant power is an upstream country. The key research question was: To what extent do power asymmetries prevent or inhibit cooperation between riparian states over water resources? This was linked to the question of how institutions help to mitigate competition for natural resources and how states interact in multilateral settings.

This book's main objective has been empirical, aiming at better understanding the selected case studies and providing a three-level analysis across three case studies. As a result, this chapter's main goal is to merge the research framework with the extended empirical findings. It begins by revisiting the dilemma and importance of water resource management. The following sections highlight the general findings and lessons learned from the case studies and will be divided in three sections – the regional framework, the River Basin Organisations (RBOs) and the micro-level cases – in order to draw a comparison and conclusions.

## Water Resource Management Revisited

Water resource management is important globally and is an issue in all regions of the world. Transboundary water relations are complex and all basins are different because water resources allocation, terrain and, even more importantly, institutional infrastructure are different. Similarly, the nature of power relations is unique in each river basin as highlighted in the empirical work. Overall, conflict between and within states over freshwater resources poses an interesting challenge for scholars and policymakers due to competing claims over water and the concept of territorial sovereignty that is challenged by transnational rivers.

Overall, there are 261 transboundary river and lake basins worldwide, which cover nearly half of the earth's land surface. This accounts for approximately 60 percent of the global freshwater flow, and over 90 per cent of the world's population lives in a country that shares a river basin.[1] Water deserves special attention because of its three key characteristics: it has no substitute; it cannot be secured in sufficiently large quantities through long-distance trade deals; and due to the interconnectivity of the hydrological system, the actions of one country in its water management have a direct bearing on the interests of neighbouring countries.[2]

While freshwater is fundamental to all ecological and societal activities, such as energy and food production, industrial development, transportation, health and employment, it is often unequally distributed within a region or even within one country. As a result, future developments of many countries depend on the successful management and allocation of these transnational resources, and, in particular, the successful cooperation between riparian states regarding transboundary water management. Some key problems related to water management include water shortages, inefficient water use, inadequate sanitation, destruction of the ecosystem, water logging and salinisation, and water pollution or flooding. Not all of these major problems are relevant in all transnational river basins; challenges vary. For instance, in the past, the Danube, Rhine and La Plata rivers experienced major problems with water pollution, whereas in the Euphrates River water quantity was and still is a key challenge between the riparian states.[3] On the Mekong River, water governance is closely related to environmental effects, especially regarding water quality and water flow. For instance, damming on the upstream river not only reduces the water flow aggravating the problem of water scarcity or allocation in the downstream countries, but also influences the water and soil quality affecting interdependent ecosystems such as the Tonle Sap Lake in Cambodia or the Pantanal wetlands in the La Plata Basin.

These issues are also interlinked with growing energy demands, an important aspect that can become a problem when hydropower developments conflict with environmental issues. As a result, water resource management and allocation play a crucial part in potential conflict between riparian states in most, if not all, regions in the world.

## Overview of Findings and Lessons

Based on my empirical findings, water resource management can only be understood when various aspects of the challenges, all levels of involvement and the various relevant actors are included. Moreover, in order to understand water resource management and its effectiveness, it is important to recapitulate the different levels of governance and their importance. At a global level, there are two main international conventions that aim to establish general principles for the use of transboundary water resources, including the United Nations Economic Commission for Europe (UNECE) Water Convention and the United Nations (UN) Convention on the Law of the Non-Navigational Uses of International Waterways.

The 1992 UNECE Water Convention, also called the Helsinki Convention, was originally negotiated as a regional instrument in Europe; however, the convention was amended in 2003 to allow accession by all UN member states.[4] Amendments effective from February 2013 turned the convention into a global legal framework for transboundary water cooperation, and non-Economic Commission for Europe (ECE) countries could join the convention from late 2015.[5] However, the UN Water Convention does not include a supranational element or the possibility of sanctions and is limited to the codification of international customary rights. As a result, the success of the convention as a global framework is yet to be seen.

The second global convention is the 1997 UN Convention on the Law of the Non-Navigational Uses of International Watercourses. This convention recently entered into force in August 2014 with the 35th country ratifying, accepting or approving it.[6] However, delays in the ratification process in the last 15 years show that the management of transboundary water courses is not a priority area in the world's political agenda and that riparians, so far, prefer to have bilateral or multilateral negotiation between riparian states instead of ratifying a global convention.

Of the three empirical case studies used in this study, only few riparians out of 30 signed the UN Water convention or the UN Convention on Non-Navigational Uses. This confirms the view that in regards to transboundary water resource management, RBOs and regional frameworks are more important and influential than the current global frameworks. The regional level plays an especially important role in developing countries, as conflicts are more likely to occur on local or regional levels, since common property resources may be more critical to survival and less easy to replace or supplement.[7]

### *Regional Perspective – European Union (EU), Association of Southeast Asian Nations (ASEAN) and Southern Cone Common Market (Mercosur)*

Looking through this regional prism allows a methodical approach that combines the micro level of domestic affairs and states with the middle level of state interactions and the macro level of the international system.[8] This demonstrates the interlinkages of all three levels and the influence they have on each other, something that is especially important in a transnational issue area. Overall, the regional level gained importance throughout the 1990s because of the worldwide globalisation process and a stronger interconnection between states and regions.[9] This is also supported by Barry Buzan, who argues that the collapse of the Cold War order and the subsequent decentralisation of the international system strengthened regionalism.[10]

As a result, this approach also justifies why certain regions are compared and what impact the regions have on the local and global levels.[11] This also links to the regional security complex which is 'a set of units whose major processes of securitization, desecuritization, or both, are so interlinked that their security problems cannot reasonably be analysed or resolved apart from one another'.[12] Linking this to the three case studies and the three regions, the importance of the regional framework for natural resource management is highlighted. Regionalism has gained in significance and is currently the focus of both water resource management theory and policy.

As highlighted in Chapter 3, ASEAN is based on the traditional regional practices of consensus-finding and consultation, which are unique not only to ASEAN but also to its subregions and multilateral cooperation.[13] More specifically, in regards to environmental cooperation among its member states, ASEAN introduced the ASEAN Expert Group on the Environment in 1978. Environmental cooperation in ASEAN can be divided into three phases, which started with a non-formal web of soft law declarations, resolutions and action plans avoiding non-binding agreements; followed by an emphasis on environmental protection, collective responses and commonalities; and later expanding to subsidiary bodies including the Environmental Ministers meeting and the ASEAN Secretariat.[14]

Early key focus areas included the management of shared natural resources regarding biodiversity and pollution control, mainly referring to haze pollution through forest fires to acquire agricultural land. Currently, ASEAN environmental cooperation includes water management through the ASEAN Cooperation on Environment. The environmental challenges faced by ASEAN are linked to its culture and principles of non-interference and non-binding decision-making. Other problems include the general lack of concrete instruments that are able to translate the regional commitments on an ASEAN level into national policy and its implementation.[15] To summarise, while ASEAN is an umbrella for the subregion, environmental matters and guidelines do not influence the subregion strongly as these procedures and guidelines are often implemented slowly and not monitored thoroughly.

Chapter 4 highlighted that the EU has played a significant role in regards to developments along the Danube River and impacted on most of the riparian countries that are either members, candidate countries or included in the EU Neighbourhood Policies. In general, EU expansion as well as EU enlargement and integration policies – including environmental legislation – has had a big impact on the Danube and the International Commission for the Protection of the Danube River (ICPDR) itself. During the creation of the EU and in its first decades, environmental matters were not dominant, and if they were enacted at all it was mainly due to concerns about trade and competition, as well as the impact of pollution on human health.[16] However, this changed over time with better scientific understanding of environmental problems, the increase of public interest and headlines of environmental disasters such as the Sandoz chemical spill polluting the Rhine River in 1986.

As a result, new laws were implemented, including standards for protection against hazardous chemicals and the introduction of sustainable development in the 1990s.[17] This was further developed into the Water Framework Directive (WFD), which came into force in 2000, covering the protection of inland surface waters, transitional waters, coastal waters and groundwater. The WFD also established the River Basin Management approach, viewing the water systems as a single geographical and hydrological unit and introducing the idea that every river basin in the EU needs to provide a River Basin Management Plan.[18]

Another important step of the WFD, which had a significant influence on the riparian states, was the role of public participation, which means that during the WFD process all interested parties such as water users, Non-Governmental Organisations (NGOs), local and regional authorities, industries and agriculture were involved. This notion of public participation was also supported by the River Basin Management Plan as can be seen in the Danube case, where there were several observers commenting on progress and initiatives. Overall, EU development and enlargement, especially integration policies in 2004 and 2007, are directly linked to the riparian countries in the Danube River, and EU legislation has had a large impact on the basin. Today, the influence of the EU in regards to environmental legislation and awareness places considerable responsibility on the existing and future EU countries and provides new challenges and opportunities for cooperation on the Danube.

As demonstrated in Chapter 5, transformations in the 1970s and 1980s, including the Brazilian–Argentinean relationship, the decline of United States (US) hegemony and the struggle for natural resources resulted in the creation of Mercosur in the 1990s. A key aim of the treaty was the idea of 'economic development with social justice'.[19] Mercosur's focus was on trade, including the goal of a common market among four countries with the free circulation of capital, services, goods and workers from 1995. This also included the creation of a common external tariff; a common trade policy towards third countries; and the coordination of macroeconomic policies.

Environmental matters were included in Mercosur in 1995 in a subgroup, but it is important to stress that this focused only on the inclusion of environmental matters in large development and infrastructure projects, similar to the Greater Mekong Subregion (GMS) in the Mekong region, and not on environmental issues and challenges in general. Overall, Mercosur has been based on a state-led approach, not driven by the private sector or economic interdependence.[20]

In general, South America has been using formal regional trade treaties as the main channel of integration. Regional integration in South America includes a more mixed picture than Southeast Asian regional development. Mercosur, for instance, includes characteristics of 'old' and 'new' regionalism. These include state-led, mainly commercial and political initiatives with little influence of civil society organisation but also an open and inclusive approach.[21] Open regionalism is also a common feature within Southeast Asian regional development.

Although the structures of Mercosur and the Commission of the EU are similar, Mercosur's common bodies do not include enforcement tools. As a result, it cannot make member states comply with the common market rules. In general, the common market reaffirms the dependency of the participant countries on Brazil and Mercosur does not make any financial transfers to develop or diversify the other economies, as distinct from the EU system that supports less developed countries.

There are similarities between the EU in the 1960s and Mercosur in 1990s. However, in 2000 Europe continued to progress, whereas Mercosur stagnated and its current prospects are rather negative due to a macroeconomic imbalance in Argentina, protectionism and the entrance of Venezuela into Mercosur. As a result, the less institutionalised option of Mercosur, without being a supranational institution, makes it a very different case to the EU.[22] Overall, integration has occurred under Mercosur and there is economic, cultural and social interdependence. Mercosur improved the process of cooperation in the region; however, it should never have been a customary union.

### *Regional Framework – Summary*

European enlargement and integration plays a crucial part in the developments along the Danube River and is a key part of further successful cooperation. Overall, the Danube is a good example to show how regional integration, specifically European integration, can encourage cooperation along the river. The Danube plays an important part in European integration and future development. Equally, the integration process plays a pivotal role for positive cooperation between the riparian countries on the Danube River. While ASEAN is an umbrella for the subregion, environmental matters and guidelines do not influence the subregion strongly – as it is the case, for instance, on the Danube River with the binding EU WFD – given that the procedures and guidelines are often slowly implemented and are not adequately monitored.

Moreover, non-interference and non-binding agreements impact on the member countries differently. One of the key challenges is the implementation from regional frameworks into national policies and law. In comparison to the Mekong and Danube River Basin, cooperation on the La Plata River has focused on bilateral treaties and agreements instead of multilateral ones. The agreements and treaties were overall successful in regards to economic development for all of the five riparians; however, the agreements lacked, for several decades, the inclusion of environmental issues such as deforestation or soil degradation, which is now attempted through the new Strategic Action Plan focusing on a basin-wide integrated water resource management approach.

Overall, all three regional frameworks are different. ASEAN is related to trade, and national views and territorial claims are still dominant. As a result, ASEAN is a rather loose and less focused institution with no ambition to become anything more. The EU is more than a customary union and has a strong institutional base impacting on all member countries. Mercosur is located somewhere in the middle between both approaches and previously had ambitions to become a bigger and stronger institution, but today is less cooperative than it was anticipated. These findings confirm my hypothesis that cooperation is more likely if a sense of regionalism exists.

## ***River Basin Organisations (RBOs) – Similarities, Differences and Lessons***

Besides understanding the regional framework and its importance for effective water resource management, the importance of RBOs and their function in this research has been highlighted. This is especially important as transboundary water resource management and the role of cooperation and conflict are ongoing challenges, and institutions play a major part in addressing them.

As established earlier, this focus acknowledges that there are different levels of cooperation, which can range from notification processes and data exchanges to more sophisticated levels, including institutional agreements. As a result, having a functional and effective organisation is important, because millions of people live on the river banks and the subregions are crucial for the economic development of the riparian states. Moreover, intergovernmental institutions can reduce power asymmetry between states by regulating reciprocity and cooperation.[23] Further, as highlighted in the detailed empirical work earlier, there exist different characteristics and functions within the existing RBOs that vary from basin to basin, being affected by the regional framework, national policies and the present water resource management challenges. The structure, funding, agenda and acceptance of RBOs within the riparian countries play a key role in dealing successfully and, over the long term, with the challenges of water governance in transboundary river basins.

### *RBOs – Membership and Funding*

The Mekong River Commission (MRC) includes the four Lower Mekong countries and excludes the two upstream countries China and Myanmar. Already since the early 1990s, the MRC secretary stressed the importance of including the upstream countries, and as a result both countries became dialogue partners of the MRC in 1996 and are slowly expanding their (non-binding) participation in its various forms, including data exchange and scientific research. However, as only dialogue partners, China and Myanmar do not need to follow the decision-making of the MRC and do not need to commit themselves to any agreements or decisions. At present it is unthinkable that either country would join the MRC.[24] Further problematic is that in 2011 around 80 per cent of the MRC funding came from external donors, including Australia, Sweden, Germany and the Asian Development Bank (ADB), whereas less than 20 per cent came from the riparian countries themselves.[25] Overall, the MRC is an international organisation between the four downstream countries in which the mainly foreign donors play a crucial role; this is seen as a problem because of the very low local ownership and acceptance. However, this might change in the future due to a major restructuring of the MRC which started in 2016. This also includes the appointment of the first riparian CEO of the MRC Secretariat Pham Tuan Phan from Vietnam.

The ICPDR includes all riparian states with a share of over 2,000 square kilometres of the Danube River and the European Union. Additional countries on the Danube – not members of the ICPDR – include EU member states Italy and Poland and non-EU members Switzerland, Albania and Montenegro. All five countries have only minor shares on the Danube and no interest in the water governance of it. Additionally, they still have ties with the ICPDR countries through the EU WFD and through bilateral ties with the other non-EU countries. As a result, in contrast to the MRC, the Danube brings together all the relevant riparians with major shares and the excluded riparians are not a hindrance for an effective RBO. Besides the riparian states, the ICPDR includes the involvement of other active non-state actors with observer status, including NGOs, organisations representing private industry and intergovernmental organisations.

The ICPDR is funded by the riparian member countries equally, which means that all the countries provide the same amount of money to run the ICPDR secretary, independent of their Gross Domestic Product (GDP) and size (although four riparians currently pay less due to financial constraints).[26] Besides funding the Secretariat, the riparian states pay for the country's specific personal costs, meeting settings, organisation and other costs involved in the participation at working groups or annual meetings.

The Comité Intergubernamental Coordinador de los Países de la Cuenca del Plata (CIC) includes all the five riparian states of the La Plata River Basin. Most riparians are also simultaneously involved in other bilateral or multilateral institutions dealing with specific issues on the La Plata, which include, for instance, the Itaipu Bi-National Entity or the treaty on the La Plata River and its Maritime Front. In 2014, the total budget for the CIC was low in comparison to the funding of the ICPDR or the MRC. The CIC is paid based on GDP and size of the riparians, with Brazil and Argentina paying around 30 per cent, Uruguay a bit less than 20 per cent and Paraguay and Bolivia a bit above 10 per cent.[27] Additional funding is available for different programs and projects.

To sum up, the membership of the institution is important as most river basins – and all of the three case studies – try to establish a basin-wide holistic approach. This puts even more emphasis on the participation of all riparians. As the Mekong case study shows, problems can arise if upstream countries are not involved, and this weakens the role of the institution. Similarly, it is important that all countries have equal standing in the institutions and are not dominated by one or a few countries.

### *Institutional Design*

Besides the funding and membership of the RBO, the institutional design and structure plays a major role in the success and adaption of an institution. The MRC today consists of three permanent bodies – the Council, the Joint Committee (JC) and the Secretariat.[28] The Ministerial council meeting is annual and one member of each country meets at the ministerial or cabinet level. The JC of senior bureaucrats provides the organisation board with technical and logistical support. The Secretariat, led by the Chief Executive Officer, plays a crucial role in the direction and focus of the organisation. Often the atmosphere and orientation of the MRC appears to fluctuate based on arduous issues and the political situation in the member countries or based on the leadership as highlighted in Chapter 3. This can result in inconsistent policies under different Chief Executive Officers (CEOs) and in controversy and difficulties within the organisation.[29]

The ICPDR is divided into the Permanent Secretary and the President. Technical and administrative staffs of the Secretariat – which are minimal in comparison to the MRC – supervise and control the quality of the Secretariat's main functions and tasks, including project staff.[30] Besides the Permanent Secretary, the ICPDR is officially represented by the President. The Presidency of the Commission changes in alphabetical order among the riparian countries annually and its influence varies a lot, with input ranging from a low contribution and interest to a very high importance and involvement of the President.

The members of the ICPDR and their representatives meet annually at the end of the year and also have an additional meeting during the year. Besides this, there exist other bodies of the ICPDR, including the Expert Groups and Task Groups. The working groups often have a regular number of people to prepare the proposals for the annual meeting. These are a key part of the ICPDR's work. The decisions in the working groups and the annual meeting are generally accepted with consensus. The ICPDR has a small number of staff and a narrow focus (water quality, environmental protection and flood management) but is effective; whereas the MRC is a rather broad organisation that lacks focus and effectiveness.

The CIC includes, besides the President and the Project Unit, the permanent General Secretary. The Secretariat of the CIC is responsible for the administration, communication and coordination of activities and projects between the member states.[31] The CIC includes two representatives from each country, one political and the other a technical representative, who meet twice a year. Since its creation the treaty was adjusted twice, in 1992 and 2001, to include technical experts and to allow meetings of representatives outside the annual meetings in order to accelerate decision-making on administrative issues.

Besides the ministers' annual meeting, extraordinary meetings can be held if necessary. Annual meetings at the Foreign Ministers' level started in 1969 when the institutional framework was established. In general, deliberation and decision-making is based on the unanimity rule, which sometimes blocks the processes as the affirmative vote of all five riparians is necessary in order to adopt a decision and no country can be absent.[32] The General Secretary of the CIC is elected for a two-year term and can be re-elected for two additional terms of one year each.[33] The CIC staff is – in comparison to the political representatives – relatively stable and the number is small, similar to the number of staff at the ICPDR. Overall, the CIC is seen as an advisory body that can recommend but cannot require actions from the member states, which causes a severe limitation of its power in a similar way to water management bodies on other river basins, including the MRC.[34]

To sum up, it is important to have a functional secretariat with technical expertise but also political support through political representatives. A political representative is important to guarantee the implementation of projects and avoid delays, ongoing changes and interruptions. Additionally, a technical representative with the necessary expertise and knowledge is important for the practical day-to-day work.

*Public Participation and Non-State Actors*

Besides the membership, funding and institutional design, the role of public participation and non-state actors will be addressed briefly as these actors are increasingly influencing the agenda of institutions and the institutions themselves. As highlighted in Chapter 3, public participation and non-state actors play a less dominant role in the Mekong Basin. For instance, civil society and NGO networks were active in Cambodia (including the River Coalition in Cambodia or the NGO Forum) advocating their standpoint towards the Lower Sesan II Dam and trying to influence the Cambodian government. However, often this is unsuccessful because their influence can be very limited and not all actors are able to agree on a standpoint or achieve a common consensus – thus reducing the possibility of positive outcomes.[35] Different views make it difficult to voice one opinion, which is important in order to achieve results in a context where civil society already has a difficult standpoint and public participation is often limited or non-existing.

NGOs play a crucial part in the Danube Basin, although they are established differently depending on the country. The ICPDR includes the involvement of other active non-state actors with observer status, including NGOs, private industry and intergovernmental organisations. All observers are invited to the meetings and they can provide comments and inputs to the proposals and raise awareness for their projects and goals in regards to the Danube River. This is one of the key reasons for the successful work of the ICPDR and it raises the necessary public awareness to support the developments on the Danube. It often also achieves a balance between different ideas and opinions, due to the representation of a wide diversity of actors involved in the process. This is particularly important in regards to Environmental Impact Assessments (EIAs) of large projects where NGO participation besides other activist groups is legally necessary and should play an active role.

Public participation varies in the La Plata River between the riparians, and the focal points are different. However, the pulp mill incident showed that the public can have an impact on decisions and is able to delay processes. The overall influence of the public on institutions is increasing. It has been evolving over time: we could see little resistance in earlier cases such as the Itaipu Dam, whereas there is stronger involvement in more recent cases.

The case of the ICPDR shows that a public participation process can be effective and supportive for an institution and its agenda in the long term. The La Plata case shows that opposition exists but is not influencing the developments as much as on the Danube. Similarly, in the Mekong the Procedures for Notification, Prior Consultation and Agreement (PNPCA) process shows that there is still a lack of public participation regarding large projects, and reliable information is often unavailable.

### *RBOs – Key Findings and Lessons*

This section highlighted that institutions play a key role in achieving water resource management on transboundary rivers. The effectiveness and functionality as well as the adaptability of institutions play a major role, as was highlighted in the three RBO case studies focusing on the importance of the institutional structure, funding, membership and the role of the CEO and Secretariat.

Key findings of the book include the importance of institutions being steered by riparian countries, including funding from the countries themselves, not just from outsiders, as this improves the responsibility for the institution and its legitimacy. Also, if the RBO is funded internally, the riparians are accountable for the actions and outcomes impacting on its effectiveness. Equal funding is preferable; however, it is often not viable due to the different economic standards of the riparians. However, possible solutions include temporary exceptions or reductions, as in the case of the ICPDR, or different percentage payments according to GDP and population, as it is the case of the CIC. Nonetheless, settling on the payments the poorest country can pay can lead to a financial budget which is too small and not sufficient to support a functional institution.

This is also true for the representative of the secretariat as a riparian representative provides a local representation. However, sometimes it is difficult to agree on one person who is 'impartial'. This can also be problematic in regards to staff in the secretariat, which is the case in the MRC where national staff is linked to a quota and the staff is delegated by the country's leaders. This is true also with political representatives in the case of the CIC, who change frequently. Continuity of staff is a key point in guaranteeing the effectiveness of the institution and an ongoing progress without interruptions. This includes a secretariat with technical expertise as well as political support through political representation, as this is the key to efficient work on specific issues.

Moreover, the membership of the institution is also crucial as most river basins try to establish a basin-wide approach. This places even more emphasis on the participation of all riparians. As the Mekong case study shows, problems can arise if upstream countries are not involved as this weakens the credibility of the institution. Also, the case of the ICPDR shows that a public participation process can be effective and support the institution and its agenda in the long term. Importantly, it is essential to remember that these conclusions are not only applicable to RBOs but also to other institutions in multilateral settings dealing with transnational challenges.

## ***Case Studies – Findings and Lessons***

At the third level of analysis, the research included extended in-depth material on three micro-level case studies within the three river basins, covering an extended time frame to showcase what is actually happening on the ground and how this informs theory and rhetoric at the institutional and regional levels.

Although circumstances, challenges and problems vary from river basin to river basin, the research covered – by presenting micro-level case studies in three broad areas – hydropower development, water pollution through waste spills and challenges through navigation projects. Hydropower developments on the Mekong included the Xayaburi Dam in Laos and the Lower Sesan II on a major tributary in Cambodia; the Gabcikovo–Nagymaros cascade on the Danube River between Hungary and Slovakia; and the Itaipu Dam on the Parana River in the La Plata River Basin. Water pollution through waste spills included the Baia Mare incident in Romania and the pulp mill incident on the Uruguay River. Navigational projects included the controversial Hidrovia navigation channel on the Paraguay River, the Bystroe Canal in the Ukraine and the navigation project on the Upper Mekong River.

For instance, the Itaipu Dam on the Parana River and the Gabcikovo Dam project on the Danube both had an impact on the riparians along the basin and resulted in disputes. The Gabcikovo case was submitted to the International Court of Justice (ICJ) and is still unresolved, although no longer a pressing issue. Similarly, the pulp mill incident could not be solved between Argentina and Uruguay; it involved first Mercosur and later the ICJ. The ICJ was used as a mediator and decided the case; however, implications for both countries still exist and so does the impact on their relations. Furthermore, in comparison to the Gabcikovo project and the Bystroe Canal on the Danube, decisions on the Mekong River such as the Xayaburi Dam went ahead mostly due to the support of the political elite and without sufficient and effective public participation.[36] Analogous to the key decision on the Mekong River, the Gabcikovo project construction went ahead mostly due to the support of the ruling elite and Slovakia's nationalist rhetoric focusing on industrialisation and national development. Natural resource management and river cooperation are used as political instruments and power displays; for instance, after the Gabcikovo withdrawal of Hungary, which was linked to the revolution in the country.

The Baia Mare incident and the Bystroe Canal project showed the increasing importance of NGOs and the public, who can influence countries' decisions and provide awareness within the community. However, this influence is still limited in some of the less developed Eastern Europe countries and in the Mekong and La Plata Region. It also varies amongst the riparians as was highlighted between Brazil and Paraguay for instance, in regards to environmental matters on the La Plata. The case studies covering a broad range of time frames also showed that things develop and evolve over time; RBOs can learn from each other and adapt. For instance, the CIC is currently involved in a major project, funded by the Global Environment Facility (GEF), which was an essential part of the development of the ICPDR and the current framework on the Danube. The micro-level analysis also highlighted that the role of the dominant states varies within the basins. Whereas Germany and Austria as the most upstream countries on the Danube and Brazil on the La Plata River are included in the RBOs, China is not part of the MRC but is involved as an observer and in the GMS.

Overall, the micro-level case studies have demonstrated that the geographical location and the power of the riparians have an influence on the institution and the decision-making process. However, this is limited and not as severe as expected, which leads to the conclusion that RBOs are effective in achieving a more balanced cooperation between asymmetrical actors and players. This, in turn, will have a positive impact on the effectiveness of the institution, which again links to the regional framework and can influence all three levels of analysis.

## Notes

1 Task Force on Transboundary Waters, "Transboundary Waters: Sharing Benefits, Sharing Responsibilities," UNESCO, Zaragoza, Spain, 2008.
2 Loh Su Hsing, "China Holds the Key to Asia's 'Blue Gold'," *Jakarta Globe*, 15.12.2011.
3 Miriam R. Lowi, "Water and Conflict in the Middle East and South Asia: Are Environmental Issues and Security Issues Linked?," *Journal of Environment and Development* 8, no. 4 (1999): 378; Natalie Kempkey, Margaret Pinard, Victor Pochat, and Ariel Dinar, "Negotiations over Water and Other Natural Resources in the La Plata River Basin: A Model for Other Transboundary Basins?," *International Negotiation* 14, no. 2 (2009): 260.
4 United Nations (UN), "Convention on the Protection and Use of Transboundary Watercourses and International Lakes," UN, Helsinki, 1992.
5 Ibid.
6 UN, "Convention on the Law of the Non-Navigational Uses of International Watercourses," https://treaties.un.org/Pages/ViewDetails.aspx?src=IND&mtdsg_no=XXVII-12&chapter=27&lang=en (accessed 13.10.2014).
7 Peter H. Gleick, "Water and Conflict: Fresh Water Resources and International Security," *International Security* 18, no. 1 (1993): 83.
8 Barry Buzan, *People, States and Fear*, 2nd ed. (New York: Harvester Wheatsheaf, 1991), 222.
9 Jens-Uwe Wunderlich, "Regionalism and Integration Theory. The First Wave: Traditional Approaches," in *Regionalism, Globalisation and International Order*, ed. Jens-Uwe Wunderlich (Hampshire and Burlington: Ashgate, 2007), 2.
10 Barry Buzan, "The Interdependence of Security and Economic Issues in the New World Order," in *Political Economy and the Changing Global Order*, eds. Richard Stubbs and Geoffrey Underhill (New York: St Martin's Press, 1994), 94–5.
11 Buzan, *People, States and Fear*, 186.
12 Barry Buzan, Ole Waever, and Jaap De Wilde, *Security: A New Framework for Analysis* (Boulder, CO: Lynne Rienner, 1998), 201.
13 Amitav Acharya, "Ideas, Identity, and Institution Building: From the 'ASEAN Way' to the 'Asia–Pacific Way'?," *The Pacific Review* 10, no. 3 (1997): 329.
14 Lorraine Elliott, "ASEAN and Environmental Cooperation: Norms, Interests and Identity," *The Pacific Review* 16, no. 1 (2003): 36, 38.
15 Ibid.
16 Miranda Schreurs, "Environmental Protection in an Expanding European Community: Lessons from Past Accessions," *Environmental Politics* 13, no. 1 (2004): 27.
17 John McCormick, *Understanding the European Union – A Concise Introduction*, 3rd ed. (Hampshire: Palgrave Macmillan, 2005).
18 European Commission, "Introduction to the New EU Water Framework Directive," http://ec.europa.eu/environment/water/water-framework/info/intro_en.htm (accessed 15.06.2012).

19 Michael Mecham, "Mercosur: A Failing Development Project?," *International Affairs* 79, no. 2 (2003): 369.
20 Mauricio Baquero-Herrera, "Latin America Update: Something Is Going on in Latin America," *Law and Business Review of the Americas* 12, no. 4 (2006): 613–22; Laura Gomez-Mera, "How 'New' Is the 'New Regionalism' in the Americas? The Case of Mercosur," *Journal of International Relations and Development* 11, no. 3 (2008): 282.
21 Gomez-Mera, "How 'New' Is the 'New Regionalism' in the Americas? The Case of Mercosur."
22 Mario E. Carranza, "Can Mercosur Survive? Domestic and International Constraints on Mercosur," *Latin American Politics and Society* 45, no. 2 (2003): 69.
23 John Ikenberry and Michael Mastanduno, "International Relations Theory and the Search for Regional Stability," in *International Relations Theory and the Asia–Pacific*, eds. John Ikenberry and Michael Mastanduno (New York: Columbia University Press, 2003), 2–5.
24 Katri Mehtonen, Marko Keskinen, and Olli Varis, "The Mekong: IWRM and Institutions," in *Management of Transboundary Rivers and Lakes*, eds. Olli Varis, Asti K. Biswas and Cecilia Tortajada (Berlin: Springer, 2008), 210.
25 Mekong River Commission (MRC), "Mekong River Commission – Audited Financial Statements," MRC, Hanoi, 2012.
26 Interview with an ICPDR senior expert on the 26.04.2012 in Vienna, Austria.
27 Interview with a political representative to the CIC in Buenos Aires on the 22.04.2014.
28 MRC, "Annual Report 2001 – Mekong River Commission," MRC, Vientiane, 2002.
29 Oliver Hensengerth, "Vietnam's Security Objectives in Mekong Basin Governance," *Journal of Vietnamese Studies* 3, no. 2 (2008): 103.
30 International Commission for the Protection of the Danube River (ICPDR), "Permanent Secretariat of the ICPDR," www.icpdr.org/icpdr-pages/secretariat.htm (accessed 04.06.2012).
31 Lilian Del Castillo, "The La Plata Basin System against the Background of Other Basin Organizations," *International Journal of Water Resources Development* 27, no. 3 (2011): 511–37; Comité Intergubernamental Coordinador de los Países de la Cuenca del Plata (CIC), "El Tratado De La Cuenca Del Plata [Treaty of the La Plata Basin]," CIC, Brasilia, 1969.
32 Del Castillo, "The La Plata Basin System against the Background of Other Basin Organizations."
33 Interview with a senior expert on the politics of the La Plata River in Buenos Aires on the 24.04.2014.
34 Patrick Gilman, Victor Pochat, and Ariel Dinar, "Whiter La Plata? Assessing the State of Transboundary Water Resource Cooperation in the Basin," *Natural Resources Forum* 32, no. 3 (2008): 205.
35 Interview with an expert of the River Coalition Cambodia in Phnom Penh on the 06.12.2012.
36 Eugene K. Balon and Juraj Holcik, "Gabcikovo River Barrage System: The Ecological Disaster and Economic Calamity for the Inland Delta of the Middle Danube," *Environmental Biology of Fishes* 54, no. 1 (1999): 8.

# 7 Conclusion

This book began with the observation that cooperation and conflict in transboundary river basins should not necessarily be seen as distinct from each other as both can exist in a river basin at the same time. There are nuances between cooperation and conflict in transboundary river basins today, and these can change over time. For example, a river basin can be cooperative in some areas and less so in other areas, which was demonstrated in the empirical work of this book. Cooperation can happen if all the riparians are in a River Basin Organisation (RBO). However, cooperation can also occur if riparians work together outside of the institutions; for instance, in regards to data sharing or cooperation in specific areas only. The key research question was linked to the questions of how institutions help to mitigate competition for natural resources and how states interact in multilateral settings.

Significantly, the book has argued that both cooperation and conflict can coexist. The research has also highlighted that too often in international relations excessive focus is given to military and economic power, and that there is an insufficient understanding of what constitutes other types of power. As a result, this research has shown that power from a realist perspective, mainly defined in military and economic capabilities, is not enough to understand relations in transnational river basins. Moreover, it is important to include the role of bargaining and issue linkage as different issues between countries are dependent rather than independent variables.

The research has highlighted the importance of the regional level, as this is often underemphasised in international relations, especially when focusing on power theories. It is important to include the regional level in order to provide a comprehensive understanding of interlinkages and interdependences between states and the international system. Looking through this regional prism allows a methodical approach that combines the micro level of domestic affairs and states with the middle level of states interactions and the macro level of great power impacts on the international system.[1] This demonstrates the interlinkage of all three levels and the influence they have on each other. This was especially important in this research as it provides a framework to understand the position of the states in relation to each other and in relation to more powerful states. These dynamics are essential in understanding water resource management, as water crosses boundaries and interaction among states is unavoidable. This approach also justifies why certain regions are compared and what impact they have on local and global levels.[2]

As a result, based on my empirical findings, water resource management can only be understood when various aspects of the challenges, all levels of involvement and the varying actors present are included. A main focus of the book was on the empirical work, especially the selected case studies that make a contribution to the broader empirical research on transnational river basins, but also help to understand other multilateral environments dealing with transnational security challenges, crossing countries and territorial borders.

This book has sought to analyse how riparian states cooperate or not in asymmetric power relations when the upstream country is dominant. The main conclusions of the research are based on the three-level analyses across three case studies. In the first level of analysis, the findings confirm Hypothesis 3 that cooperation is more likely if a sense of regionalism exists. As a result, in regards to the second level of analysis, the book expected that the International Commission for the Protection of the Danube River (ICPDR) is the most effective and functional out of the three RBOs, followed by the Comité Intergubernamental Coordinador de los Países de la Cuenca del Plata (CIC) and Mekong River Commission (MRC). Overall, the findings confirm that conflict is less likely if multilateral institutions include all riparian states and power is equally distributed (Hypothesis 1). However, the empirical work also showed that cooperation can be achieved even if not all riparians are included, for instance, through observer status and data sharing (however, this is sometimes more problematic, as demonstrated in the case of the MRC). The selected cases also highlighted that the MRC is more successful in water resource management than (for instance) the CIC, which still focuses on economic development and less on an integrated water management approach.

Linked to Hypothesis 2, the empirical work showed that conflict is less likely if the existing multilateral institutions have strong legal frameworks, including dispute resolution mechanisms. However, the cases also showed (especially the micro-level analysis) that sometimes existing mechanisms are not sufficient to deal with disputes within the framework of the institutions and that other mechanisms are necessary and used. This was, for instance, the case with the Bystroe Canal, where other international conventions such as the Espoo Convention were used as mechanisms to solve the conflict, or through the International Court of Justice (ICJ), as in the case of the Gabcikovo–Nagymaros Dam system, or the pulp mill incident on the Uruguay River. As a result of these findings, it is important that a legal mechanism and framework are established. However, this does not necessarily need to be within the RBO, as other already existing mechanisms can also be used to solve disputes.

Hypothesis 4 was confirmed through the case studies that highlighted cooperation is more likely if strong linkages between the states exist. This is independent of the power relationships and geographical positioning on the river and was demonstrated in all three levels of analysis. This showed that interaction and cooperation between states – be it through a regional or bilateral mechanism in the area of natural resource management or in another area – is beneficial for cooperation on water resource management, and achieving it for the long term can overcome challenges and changing environments.

Overall, this book concludes that cooperation in transboundary river basins is possible in asymmetric power relations and that the argument that military and economic power is the only driver behind cooperation and decision-making has limited utility. Power is important, but material power is not the only form of power that can be used by states or actors within the state. The research showed – taking into consideration theoretical concepts of realism and liberalism – that besides the power relation between the riparians, issue linkage and bargaining play a central role. The influence of the regional umbrella, including the historical ties, institutional design and political system in which the basin is based, should also be taken into account. The case study analysis showed that the geographical location and the power of the riparians have an influence on the institution and the decision-making processes. However, this influence is limited and not as intense as expected, which leads to the conclusion that RBOs are effective in achieving a more balanced cooperation between asymmetrical actors and players and that this effectiveness has a positive impact on the effectiveness of the institution, which again links to the regional framework and can influence all three levels of analysis. Further, the book showed that there are nuances between cooperation and conflict, that a river basin can be cooperative in some areas and less so in other areas and that both situations can exist in parallel and change over time.

Overall, water resource management and allocation plays a crucial role in potential conflict between riparian states in most, if not all, regions in the world. As a result, the future development of many countries depends on the successful management of these transnational resources and, in particular, on the successful cooperation among riparian states regarding transboundary water management.

## Implications and Further Research

These are important observations that demand further scholarly enquiry. Deeper analysis across regions and across a variety of case studies would enhance both scholarly and policy understanding. Overall, this research provides insights about why it is important to focus on different levels of analysis and why to link them together. The research has also demonstrated how to achieve robust institutions that are able to adapt to changing environments and challenges.

Overall, the developed characteristics and features of multilateral institutions are important in RBOs but are also transferable to other institutions dealing with similar challenges, changing tasks and varying environments. Transnational threats demand cooperation, and effective mechanisms and institutions are necessary in order to be successful in tackling these threats as they become the problem of policymakers, practitioners and scholars alike as they cross borders and challenge territorial sovereignty.

The research also confirmed that there is a need to focus more on the design attributes to reach legitimacy and effectiveness in order to achieve balanced management, as previously highlighted.[3] This is important to attain flexible and robust institutions, which can adapt to changing environments and challenges. This also applies beyond RBOs to other multilateral institutions dealing with any other transnational challenges.

Other broader key findings are that it is important to include micro-analysis case studies to be aware of the on-ground implications and how these link back to the institutional level and the regional framework. Overall, it is important to combine the findings of the three levels as each of them informs and influences the other levels and cannot be understood separately from each other. This approach also helps to put theory to good practical use and test different scenarios. The empirical work is important as it allows confirmation of or disagreement with theories or highlights where adjustment is necessary. Finally, the research shows the importance of governance mechanisms and the rise of multilateralism supporting policymakers to manage transnational challenges and adapt to contemporary challenges. These findings and recommendation can only be tested and further developed in regards to the 'newer' debates on water security and the water–energy–food nexus.

## Notes

1 Barry Buzan, *People, States and Fear*, 2nd ed. (New York: Harvester Wheatsheaf, 1991), 222.
2 Ibid., 186.
3 Ramiro Berardo and Andrea K. Gerlak, "Conflict and Cooperation Along International Rivers: Crafting a Model of Institutional Effectiveness," *Global Environmental Politics* 12, no. 1 (2012): 101–20.

# Appendix

## List of Interviews and Special Libraries

(Names of representatives withheld, as per university ethics process determination)

### Mekong River Basin: 2012 and 2013

#### *Bangkok, Thailand*

Greater Mekong Subregion Core Environmental Program (GMS CEP)
International Rivers Network
International Union for the Conservation of Nature (IUCN)
Mekong Basin Disease Surveillance (MBDS)
Mekong Energy and Ecology Network (MEE Net)
Towards Ecological Recovery and Regional Alliance/Foundation for Ecological Recovery (TERRA/FER)
United Nations Economic and Social Commission for the Asia and the Pacific (UNESCAP)
United States Agency for International Development (USAID)

#### *Chiang Mai, Thailand*

Living River Siam
Regional Centre for Social Science and Sustainable Development (RCSD)
Unit for Social and Environmental Research, Chiang Mai University (CMU)

#### *Phnom Penh, Cambodia*

Asian Development Bank (ADB)
Cambodian Research Group
Cambodia Development Resource Institute (CDRI)
Equitable Cambodia
Deutsche Gesellschaft für Internationale Zusammenarbeit (GIZ) Land Management Programme
Non-Governmental Organisation Forum (NGOF)

Oxfam Great Britain
River Coalition of Cambodia (RCC)
Royal University of Phnom Penh

### *Vientiane, Laos*

Asian Development Bank (ADB)
CIGAR-Challenge Program on Water and Food (CPWF) representatives
Consulting Economist
Helvetas Swiss Intercooperation
International Financial Corporation (IFC)
International Fund for Agricultural Development (IFAD)
International Union for the Conservation of Nature (IUCN)
Mekong River Commission (MRC) representatives
Mekong Watch Japan
MRC–GIZ Cooperation Programme representatives
National Heritage Institute (NHI)
National University of Laos
United Nations Development Programme (UNDP)
World Wide Fund for Nature (WWF) Greater Mekong Programme

## Danube River Basin: 2013

### *Budapest, Hungary*

Hungarian National Development Agency

### *Vienna, Austria*

Chief Executive Officers (CEOs) from the International Commission for the Protection of the Danube River Basin (ICPDR)
Former President of the ICPDR
Representative of the ICPDR

Skype interview with a European Union (EU) representative
Skype interview with a WWF – Danube–Carpathian Programme (WWF–DCP) representative

## La Plata River Basin: 2014

### *Buenos Aires, Argentina*

Faculty of Latin American Studies in Social Science (FLACSO)
Fundacion Vida Silvestre Argentina
Law Faculty of the University of Buenos Aires

Representative of the Brazilian Embassy
Representative of the Paraguayan Embassy
Secretary General of the Comité Intergubernamental Coordinador de los Páises de la Cuenca del Plata (CIC)

### *Foz du Iguassu, Brazil*

Itaipu Dam representative

### *Montevideo, Uruguay*

Ministry for Transportation and Planning
Ministry of Housing, Land management and the Environment

Skype interview with a Brazilian expert on water management
Skype interview with CIC programme representative

## Special Libraries

Mekong River Commission Library in Vientiane, Laos
Library of the Cambodia Development Resource Institute (CDRI) in Phnom Penh, Cambodia
United Nations (UN) library in Bangkok, Thailand

# Bibliography

Acharya, Amitav. "Ideas, Identity, and Institution Building: From the 'ASEAN Way' to the 'Asia–Pacific Way'?." *The Pacific Review* 10, no. 3 (1997): 319–46.

—. *Constructing a Security Community in Southeast Asia: ASEAN and the Problem of Regional Order*. New York: Routledge, 2001.

Adam, Ruxandra. "Venezuela Is Officially Accepted in Mercosur." *Softpedia*, 22.07.2006.

Administrative Commission for the Uruguay River (CARU). "The River Uruguay Executive Commission." 1–38. Paysandu, Uruguay: CARU, n.d.

—. "Statute of the River Uruguay. Signed at Salto on 26 February 1975." United Nations – Treaty Series, 339–47. Salto, Uruguay: CARU, 1975.

Agência Nacional de Transportes Aquaviários (ANTAQ). "Bacia Do Paraguai – Plano Nacional De Integração Hidroviária [The Paraguay Basin – National Plan of Waterway Integration]." Brasilia: ANTAQ, 2013.

Aminian, Nathalie, K. C. Fung, and Francis Ng. "A Comparative Analysis of Trade and Economic Integration in East Asia and Latin America." *Economic Change and Restructuring* 42, no. 1 (2009): 105–37.

Anton, Donald K. "Case Concerning Pulp Mills on the River Uruguay." *Australian International Law Journal* 17 (2010): 213–23.

Armstrong, David, Lorna Lloyd, and John Redmond. *International Organizations in World Politics*. 3rd ed. New York: Palgrave Macmillan, 2004.

Asian Development Bank (ADB). "Annual Report 2010 – Volume 1." Manila, Philippines: ADB, 2010.

—. "Annual Report 2014 – Members, Capital Stock and Voting Power." Available online at www.adb.org/sites/default/files/institutional-document/158032/oi-appendix1.pdf (accessed 23.11.2015).

—. "Asian Development Bank – Greater Mekong Subregion." Available online at www.adb.org/gms/ (accessed 13.09.2011).

—. "Greater Mekong Subregion Ministers to Move to New Phase of Environmental Cooperation." Available online at www.adb.org/news/greater-mekong-subregion-ministers-move-new-phase-environmental-cooperation (accessed 26.07.2011).

—. "Myanmar." Available online at http://beta.adb.org/countries/myanmar/main (accessed 13.09.2011).

Association of Southeast Asian Nations (ASEAN). "ASEAN–China Dialogue Relations." Available online at http://asean.org/?static_post=overview-asean-china-dialogue-relations (accessed 20.09.2011).

—. "ASEAN – Treaty for Amity and Cooperation in Southeast Asia." Available online at http://asean.org/treaty-amity-cooperation-southeast-asia-indonesia-24-february-1976/ (accessed 22.09.2011).

—. "ASEAN and China Officially Launch the Establishment of the China–ASEAN Environmental Cooperation Centre." Available online at http://asean.org/asean-and-china-officially-launch-the-establishment-of-the-china-asean-environmental-cooperation-centre/ (accessed 24.09.2015).
—. "Hanoi Plan of Action." Available online at www.asean.org/?static_post=hanoi-plan-of-action (accessed 22.05.2016).
—. "Overview of the Association of Southeast Asian Nations." Available online at http://www.asean.org/asean/about-asean/overview/ (accessed 29.04.2015).
—. "Plan of Action to Implement the Joint Declaration on ASEAN – China Strategic Partnership for Peace and Prosperity." Available online at http://asean.org/?static_post=plan-of-action-to-implement-the-joint-declaration-on-asean-china-strategic-partnership-for-peace-and-prosperity (accessed 28.06.2011).
Association of Southeast Asian Nations (ASEAN) Cooperation on Environment. "Overview of ASEAN Cooperation on Environment." Available online at http://environment.asean.org/about-us-2/ (accessed 30.09.2013).
Avila, Patricia. "Uruguay Approves Metsa–Botnia Pulp Mill." Available online at www.reuters.com/article/2007/11/10/environment-metsabotnia-uruguay-dc-idUSL0941462920071110 (accessed 05.05.2014).
Balon, Eugene K., and Juraj Holcik. "Gabcikovo River Barrage System: The Ecological Disaster and Economic Calamity for the Inland Delta of the Middle Danube." *Environmental Biology of Fishes* 54, no. 1 (1999): 1–17.
Baquero-Herrera, Mauricio. "Open Regionalism in Latin America: An Appraisal." *Law and Business Review of the Americas* 11, no. 2 (2005): 139–83.
—. "Latin America Update: Something Is Going on in Latin America." *Law and Business Review of the Americas* 12, no. 4 (2006): 613–22.
Barclay, Glen. *A Very Small Insurance Policy*. Brisbane: University of Queensland Press, 1988.
Barreto, Damaris, Pamela Furrer, Auro Guerra, Monica Hernandez, Angelica Mateo, and Rodolfo Martinez. "Hidrovias Del Mercosur: De La Ilusion Al Desencanto [Waterways of Mercosur: From Illusion to Disenchantment]." Santa Fe, Argentina: National University of Litoral, 2008.
Batha, Emma. "Death of a River." *BBC News Online*, 15.02.2000. Available online at http://news.bbc.co.uk/2/hi/europe/642880.stm (accessed 5.12.2013).
Baviera, Aileen S. P. "Regionalism and Community Building in East Asia: Challenges and Opportunities." In *Advancing East Asian Regionalism*, edited by Melissa G. Curley and Nicholas Thomas, 229–46. Abingdon: Routledge, 2007.
—. "Accommodation with Hedging: Southeast Asia's Changing Perspectives Towards China." In *China's Rise – Threat or Opportunity*, edited by Herbert S. Yee, 176–90. Abingdon: Routledge, 2011.
*BBC News*. "Hungary Appeals for Ecological Help." *BBC News*, 15.02.2000. Available online at http://news.bbc.co.uk/2/hi/europe/643907.stm (accessed 5.12.2013).
—. "One Year On: Romania's Cyanide Spill." *BBC News*, 31.01.2001. Available online at http://news.bbc.co.uk/2/hi/europe/1146979.stm (accessed 6.1.2014).
—. "Pollution Is a Worldwide Problem Which Does Not Respect National Boundaries and Is Likely to Intensify as the Spread of Industrial Development Continues." *BBC News*, 13.12.2004. Available online at http://news.bbc.co.uk/2/hi/science/nature/4083331.stm#romania (accessed 5.12.2013).
—. "Hungary Demands Action over Pollution." *BBC News*, 14.03.2000. Available online at http://news.bbc.co.uk/2/hi/europe/677755.stm (accessed 5.12.2013).

*BBC Online*. "European Environmental Inequalities – Danube River." Available online at www.bbc.co.uk/scotland/education/int/geog/eei/rivers/danube/ (accessed 08.01.2012).

Beeson, Mark. "Asymmetrical Regionalism: China, Southeast Asia and Uneven Development." *East Asia* 27, no. 4 (2010): 329–43.

—, and Mark T. Berger. "The Paradoxes of Paramountcy: Regional Rivalries and the Dynamics of American Hegemony in East Asia." *Global Change, Peace and Security* 15, no. 1 (2003): 27–42.

Berardo, Ramiro, and Andrea K. Gerlak. "Conflict and Cooperation Along International Rivers: Crafting a Model of Institutional Effectiveness." *Global Environmental Politics* 12, no. 1 (2012): 101–20.

Berg, Gerald C. "Does Latin America Comprise Transnational 'Subregions'?." *The World Economy* 34, no. 2 (2011): 298–312.

Bernauer, Thomas. "Explaining Success and Failure in International River Management." *Aquatic Science* 64, no. 1 (2002): 1–19.

—, and Tobias Siegfried. "Climate Change and International Water Conflict in Central Asia." *Journal of Peace Research* 49, no. 1 (2012): 227–39.

Bernstorff, Andreas, and Judit Kanthak. "The Real Face of the Kangaroo." 1–13. Amsterdam: Greenpeace, 2000.

Biswas, Asti K. "Cooperation or Conflict in Transboundary Water Management: Case Study of South Asia." *Hydrological Sciences Journal* 56, no. 4 (2011a): 662–70.

—. "Transboundary Water Management in Latin America: Personal Reflections." *International Journal of Water Resources Development* 27, no. 3 (2011b): 423–9.

Bradley, Graham. "Brazil, Argentina Forge Closer Ties." *San Francisco Chronicle*, 30.04.1986. Available online at www.washingtonpost.com/archive/politics/1986/04/17/brazil-argentina-forge-closer-ties/c1b585c8-3d39-4f76-8bb9-7b0582619fa4/ (accessed 30.04.2014).

—. "Brazil–Argentina Brotherhood a Fragile Goal for S. America." *Houston Chronicle*, 11.05.1986. Available online at www.chron.com/news/world/article/Argentina-and-Brazil-agree-to-mediate-in-7940977.php (accessed 30.04.2014).

Breslin, Shaun, and Richard Higgott. "Studying Regions: Learning from the Old, Constructing the New." *New Political Economy* 5, no. 3 (2000): 333–52.

Brown, Chris, and Kirsten Ainley. *Understanding International Relations*. 3rd ed. New York: Palgrave Macmillan, 2005.

Buzan, Barry. *People, States and Fear*. 2nd ed. New York: Harvester Wheatsheaf, 1991.

—. "The Interdependence of Security and Economic Issues in the New World Order." In *Political Economy and the Changing Global Order*, edited by Richard Stubbs and Geoffrey Underhill, 89–102. New York: St Martin's Press, 1994.

—, Ole Waever, and Jaap De Wilde. *Security: A New Framework for Analysis*. Boulder, CO: Lynne Rienner, 1998.

Campbell, Ian C. "Chapter 1: Introduction." In *The Mekong – Biophysical Environment of an International River Basin*, edited by Ian C. Campbell, 1–11. New York: Elsevier, 2009.

Capaldo, Griselda. "The Regional Integration Process." *Environmental Policy and Law* 43, nos. 4–5 (2013): 222–5.

Carmin, Joann, and Stacy D. Vandeveer. "Enlarging EU Environments: Central and Eastern Europe from Transition to Accession." *Environmental Politics* 13, no. 1 (2004): 3–24.

Carranza, Mario E. "Can Mercosur Survive? Domestic and International Constraints on Mercosur." *Latin American Politics and Society* 45, no. 2 (2003): 67–104.

—. "Mercosur, the Global Economic Crisis, and the New Architecture of Regionalism in the Americas." 1–37. Latin American Trade Network Working Paper No. 125, 2010.
Cascao, Ana Elisa. "Ethiopia – Challenges to Egyptian Hegemony in the Nile Basin." *Water Policy* 10, no. 2 (2008): 13–28.
Center of Wisdom and Care for the La Plata Basin. "The La Plata Basin." Available online at www.itaipu.gov.py/es/medio-ambiente/centro-de-saberes-y-cuidados-socioambientales-de-la-cuenca-del-plata (accessed 04.10.2012).
Central Intelligence Agency (CIA). "Cambodia." Available online at www.cia.gov/library/publications/the-world-factbook/geos/cb.html (accessed 23.11.2015).
—. "Laos." Available online at www.cia.gov/library/publications/the-world-factbook/geos/la.html (accessed 23.11.2015).
Centro de Formación en Integración Regional (CEFIR). "Atlas De Energias Renovables Del Mercosur [Atlas of Renewable Energies in Mercosur]."Available online at www.cefir.org.uy/ (accessed 03.06.2014).
Chellany, Brahma. "Chinese Water Hegemony Is a Threat to Asia." Available online at www.bdlive.co.za/opinion/2013/03/07/chinese-water-hegemony-is-a-threat-to-asia (accessed 01.02.2014).
Children of Mekong. "Mr Jeremy Bird." Available online at www.childrenofthemekong.org/ (accessed 04.11.2011).
Cocklin, Chris, and Monique Hain. "Evaluation of the EIA for the Proposed Upper Mekong Navigation Improvement Project – Report Prepared for the MRC – Environment Program." Melbourne: Monash Environment Institute, 2001.
Cogan, Jacob K. "The 2010 Judicial Activity of the International Court of Justice." *The American Journal of International Law* 105, no. 3 (2011): 477–92.
Cogels, Olivier. "Mekong Hydropower Development Is Good." *Bangkok Post*, 09.01.2007. Available online at www.greengrants.org.cn/read.php?id=1353 (accessed 08.09.2011).
Comité Intergubernamental Coordinador de los Países de la Cuenca del Plata (CIC). "El Tratado De La Cuenca Del Plata [Treaty of the La Plata Basin]." Brasilia: CIC, 1969.
—. "Programa para a Gestao Sustentavel dos Recursos Hídricos da Bacia do Prata, considerando os Efeitos Decorrentes da Variabilidade e Mudanca do Clima [Program for the Sustainable Management of Water Resources in the Plata Basin, Considering the Arising Effects of Climate Change]." 1–82. Buenos Aires: CIC, 2011.
—. "Programa Hidrovia Paraguay–Parana [Paraguay–Parana Waterway Program]." Available online at www.cicplata.org/?id=hpp_intro (accessed 19.06.2014).
Cook, Peter. *Australia and Vietnam 1965–1972*. Melbourne: La Trobe University, 1991.
—, and Corinne Manning. *Australia's Vietnam War in History and Memory*. Melbourne: La Trobe University, 2002.
Costachie, Silviu, Ionica Soare, and Daniel Dieaconu. "The Nature of State Transborder Reterritorialization in the Lower Danube Lands." *Geopolitics, History, and International Relations* 2, no. 1 (2010): 152–7.
Cote, Sean. "Protesting Pulp: Argentina Battles Pulp Mills on the Uruguay River." *Law and Business Review of the Americas* 17, no. 4 (2011): 733–58.
Council of Europe. "Convention on the Conservation of European Wildlife and Natural Habitats – Application of the Convention – Summary of Case Files and Complaints." In *Standing Committee*. Strasbourg, 27–30 November, 2012.
Coviello, Manlio, Juan Gollan, and Miguel Perez. "Las Alianzas Publico–Privados En Energias Renovables En America Latina Y Caribe [The Private–Public Renewable

Energy Alliance in Latin America and the Caribbean].” 1–64. Santiago de Chile: CEPAL, 2012.

Da Rosa, J. Eliseo. “Economics, Politics, and Hydroelectric Power: The Parana River Basin.” *Latin American Research Review* 18, no. 3 (1983): 77–107.

Da Silva, Carolina Joana, and Pierre Girard. “New Challenges in the Management of the Brazilian Pantanal and Catchment Area.” *Wetlands Ecology and Management* 12, no. 6 (2004): 553–61.

Dahl, Robert A. “The Concept of Power.” *Behavioral Science* 2, no. 3 (1957): 201–15.

—. *Who Governs? Power and Democracy in an American City*. New Haven, CT: Yale University Press, 1961.

*Daily Fusion*. “China Becomes ‘Key Investor’ in Southeast Asia’s Hydropower.” *Daily Fusion*, 19.08.2013. Available online at www.greengrants.org.cn/read.php?id=1353 (accessed 08.09.2011).

Danube Commission. “Danube Commission – Officials of the Secretariat.” Available online at www.danubecommission.org/index.php/en_US/secretariat (accessed 18.06. 2015).

—. “Welcome to the Danube Commission Website.” Available online at www.danubecommission.org/index.php/en_US/welcome (accessed 18.06.2015).

Daoundy, Marwa. “Asymmetric Power Negotiating Water in the Euphrates and Tigris.” *International Negotiation* 14, no. 2 (2009): 359–89.

Davids, Gavin. “The Baia Mare Cyanide Spill.” Available online at http://news.in.msn.com/gallery.aspx?cp-documentid=3460600&page=7 (accessed 06.01.2014).

De Onis, Juan. “Brazil, Argentina Agree on Economic Integration Plan.” *Los Angeles Times*, 14.07.1986. Available online at http://articles.latimes.com/1986-07-14/business/fi-19334_1_economic-integration (accessed 20.04.2014).

Deets, Stephen. “Solving the Gabcikovo–Nagymaros Dam Conflict.” 1–2. Washington, DC: Wilson Center, 1998.

Del Castillo, Lilian. “The Plata Basin Institutional Framework.” In *Management of Latin American River Basins: Amazon, Plata and Sao Francisco*, edited by Asti K. Biswas, Newton V. Cordeiro, Benedito P. F. Braga and Cecilia Tortajada, 175–204. Tokyo: United Nations University Press, 1999.

—. “The La Plata Basin System against the Background of Other Basin Organizations.” *International Journal of Water Resources Development* 27, no. 3 (2011): 501, 511–37.

Deudney, Daniel. “The Case against Linking Environmental Degradation and National Security.” *Millennium – Journal of International Studies* 19, no. 3 (1990): 461–76.

Diamint, Rut. “Integración Y Seguridad: La Dialética De Los Argentinos [Integration and Security: The Dilemma of Argentinians].” In *Argentina, Brasil Y Chile: Integración Y Seguridad [Argentina, Brazil and Chile: Integration and Security]*, edited by Felix Aravena, 45–73. Santiago: Nueva Sociedad/FLASCO, 1999.

Dieter, Heribert. “Changing Patterns in Regional Governance: From Security to Political Economy?.” *The Pacific Review* 22, no. 1 (2009): 73–90.

Dinar, Shlomi. “Power Asymmetry and Negotiations in International River Basins.” *International Negotiation* 14, no. 2 (2009a): 329–60.

—. “Scarcity and Cooperation Along International Rivers.” *Global Environmental Politics* 9, no. 1 (2009b): 109–35.

—, and Ariel Dinar. “Recent Developments in the Literature on Conflict and Cooperation in International Shared Water.” *Natural Resources Journal* 43, no. 4 (2003): 1217–87.

Dombrowsky, Ines. *Conflict, Cooperation and Institutions in International Water Management*. Cheltenham: Edward Elgar, 2007.

Dutu, Mircea. "Legal Implications of the 'Bystroe' Danube–Black Sea Canal Project." 1–19. Bucharest: Ecological University of Bucharest, n.d.

Ebert, Suzanne. "Defending the Danube." Available online at www.internationalrivers.org/the-way-forward/river-revival/defending-danube (accessed 08.03.2012).

Elhance, Arun P. *Hydropolitics in the 3rd World – Cooperation and Conflict in International River Basins*. Washington, DC: United States Institute for Peace, 1999.

Elliott, Lorraine. "ASEAN and Environmental Cooperation: Norms, Interests and Identity." *The Pacific Review* 16, no. 1 (2003): 29–52.

—. "Environmental Challenges, Policy Failure and Regional Dynamics in Southeast Asia." In *Contemporary Southeast Asia*, edited by Mark Beeson, 248–65. New York: Palgrave Macmillan, 2009.

—. "ASEAN and Environmental Governance: Rethinking Networked Regionalism in Southeast Asia." *Procedia Social and Behavioural Sciences* 14 (2011): 61–4.

Emmers, Ralf, Beth Greener-Barcham, and Nicholas Thomas. "Institutional Arrangements to Counter Human Trafficking in the Asia–Pacific." *Contemporary Southeast Asia* 28, no. 3 (2006): 490–511.

*Environment News Service*. "Small Oil Leak Points up Big Danube River Issues." Available online at http://ens-newswire.com/2012/11/27/small-oil-leak-points-up-big-danube-river-issues/ (accessed 21.10.2013).

Environmental Investigation Agency. "Crossroads—The Illicit Timber Trade between Laos and Vietnam." 1–22. London: Environmental Investigation Agency, 2011.

Erskine, Brian. "Disaster on the Danube." In *Borders and Border Politics in a Globalizing World*, edited by Paul Ganster and David E. Lorey, 237–46. Oxford: Rowman and Littlefield, 2005.

European Anglers Alliance (EAA). "European Anglers Alliance – History." Available online at www.eaa-europe.org/about-eaa/about-eaa.html (accessed 22.06.2012).

European Commission. "Directive 2000/60/EC of the European Parliament and of the Council of 23 October 2000 Establishing a Framework for Community Action in the Field of Water Policy." *Official Journal of the European Communities* L327 (2000): 1–21.

—. "Joint Mission of the Expert Team of the European Commission and International Conventions to the 'Bystroe Project' in the Ukrainian Part of the Danube Delta (6–8 October)," 1–11. Odessa, Ukraine: European Commission, 2004.

—. "Commission Opinion on Montenegro's Application for Membership of the European Union," 1–13. Brussels: European Commission, 2010a.

—. "EU Strategy for the Danube Region – First Contribution by Slovenia." 1–9. Brussels: European Commission, 2010b.

—. "Conditions for Enlargement." Available online at http://ec.europa.eu/enlargement/archives/pdf/press_corner/publications/corpus_en.pdf (accessed 15.06.2012).

—. "Danube Strategy: Transport Ministers to Cooperate on Navigability." Available online at http://europa.eu/rapid/pressReleasesAction.do?reference=IP/12/579&format=HTML&aged=0&language=EN (accessed 12.06.2012).

—. "Enlargement." Available online at http://ec.europa.eu/environment/enlarg/index_en.htm (accessed 04.06.2012).

—. "Enlargement from 6 to 27 Members and Beyond." Available online at http://ec.europa.eu/enlargement/archives/pdf/press_corner/publications/corpus_en.pdf (accessed 04.06.2012).

—. "Introduction to the New EU Water Framework Directive." Available online at http://ec.europa.eu/environment/water/water-framework/info/intro_en.htm (accessed 15.06.2012).

European Union (EU). "Treaty of Lisbon." *Official Journal of the European Union*, 17.12.2007. Available online at http://eur-lex.europa.eu/legal-content/EN/TXT/?uri=celex%3A12007L%2FTXT (accessed 22.06.2012).

Feagin, Joe R., Anthony M. Orum, and Gideon Sjoberg. *A Case for the Case Study*. Chapel Hill, NC: University of North Carolina Press, 1991.

Fernquest, Jon. "Rice Exports: Fall from 1st to 3rd." *Bangkok Post*, 25.10.2012. Available online at www.bangkokpost.com/learning/learning-from-news/318215/rice-exports-falls-from-1st-to-3rd (accessed 5.1.2013).

Frey, Frederick W. "The Political Context of Conflict and Co-Operation over International River Basins." *Water International* 18, no. 1 (1993): 54–68.

Freyberg-Inan, Annette. *What Moves Man*. Albany, NY: State University of New York Press, 2004.

Fuerst, Heiko. "The Hungarian–Slovakian Conflict over the Gabcikovo–Nagymaros Dams: An Analysis." 1–15. Hamburg: University of Hamburg, n.d.

Galli, Alvaro. "La Experencia De La Hidrovia Paraguay–Parana [The Experience of the Paraguay–Parana Waterway]." In *Navegacion y Control de Inundaciones en el Rio Magdalena: Un Proyeto Pais [Navigation and Control of the Flooding in the Magdalena River: A Country Project]*. 1–12. Bogota, Colombia, September, 2011.

Garcia, Beatrix. *The Amazon from an International Law Perspective*. New York: Cambridge University Press, 2011.

Gardini, Gian Luca. "Who Invented Mercosur?." *Diplomacy and Statecraft* 18, no. 4 (2007): 805–30.

George, Alexander, and Andrew Bennett. "Case Studies and Theory Development." In *Case Studies and Theory Development in the Social Sciences*, edited by Alexander George and Andrew Bennett, 3–36. Cambridge, MA: Harvard University, 2005.

Gerlak, Andrea K. "The Global Environment Facility and Transboundary Water Resource Management: New Institutional Arrangements in the Danube River and Black Sea Region." *The Journal of Environment and Development* 13, no. 4 (2004): 400–24.

Gilman, Patrick, Victor Pochat, and Ariel Dinar. "Whiter La Plata? Assessing the State of Transboundary Water Resource Cooperation in the Basin." *Natural Resources Forum* 32, no. 3 (2008): 203–14.

Gilpin, Robert. *War and Change in World Politics*. New York: Cambridge University Press, 1981.

—. *The Political Economy of International Relations*. New York: Princeton University Press, 1987.

Glaser, Charles. "Realists as Optimists: Cooperation as Self-Help." *International Security* 19, no. 3 (1995): 50–93.

Gleick, Peter H. "Water and Conflict: Fresh Water Resources and International Security." *International Security* 18, no. 1 (1993): 79–112.

Goh, Evelyn. "China in the Mekong River Basin: The Regional Security Implications of Resource Development on the Lancang Jiang." In *Non-Traditional Security in Asia*, edited by Mely Caballero-Anthony, Ralf Emmers and Amitav Acharya, 225–46. Hampshire and Burlington: Ashgate, 2006.

—. "ASEAN Adventures." *The National Interest* 94 (2008): 51–68.

Gomez-Mera, Laura. "How 'New' Is the 'New Regionalism' in the Americas? The Case of Mercosur." *Journal of International Relations and Development* 11, no. 3 (2008): 279–308.

Gordon, Frederick D. "The 1994 Convention on Cooperation for the Protection and Sustainable Use of the River Danube (High Level of Cooperation)." In *Freshwater Resources and Interstate Cooperation*, edited by Frederick D. Gordon, 91–100. Albany, NY: State University of New York, 2008.

Gottgens, Johan F. "The Paraguay–Parana Hidrovia: Large Scale Channelization or a 'Tyranny of Small Decision'." In *The Pantanal of Brazil, Bolivia and Paraguay*, edited by Frederick A. Swarts, Chapter 14. Gouldsboro, PA: Hudson MacArthur Publishers, 2000.

—, Ronald H. Fortney, Jill E. Meyer, James E. Perry, and Brian E. Rood. "The Case of the Paraguay–Parana Waterway (Hidrovia) and its Impact on the Pantanal of Brazil: A Summary Report to the Society of Wetland Scientists." *Wetlands Bulletin* 15, no. 3 (1998): 12–18.

—, James E. Perry, Ronald H. Fortney, Jill E. Meyer, Michael Benedict, and Brian E. Rood. "The Paraguay–Parana Hidrovia: Protecting the Pantanal with Lessons from the Past." *BioScience* 51, no. 4 (2001): 301–8.

Greater Mekong Subregion (GMS) Environment Operations Centre. "Fast Facts – the Greater Mekong Subregion Core Environment Program and Biodiversity Conservation Corridor Initiative (CEP–BCI)." Bangkok: GMS, 2010.

Greater Mekong Subregion (GMS) Secretariat. "Greater Mekong Subregion." Available online at www.adb.org/gms (accessed 17.10.2011).

Grieco, Joseph M. "Anarchy and the Limits of Cooperation: A Realist Critique of the Newest Liberal Institutionalism." *International Organization* 42, no. 3 (1988): 485–507.

Guercan, Efe Can. "New Regionalism and Radical Identity Formation in Latin America: Towards an 'Alter–Global' Paradigm." *Journal of Social Research and Policy* 1, no. 2 (2010): 19–33.

Guti, Gabor. "Ecological Impacts of the Gabcikovo River Barrage System with Special Reference to Umbra Krameri Walbaum, 1792, in the Szigetköz Floodplain." In *Annalen des Naturhistorischen Museum Wien, Serie B fuer Botanik und Zoologie [Annals of the Museum of Natural History Vienna, Series B for Botany and Zoology]*. 466–9. Vienna: Naturhistorisches Museum Wien [Museum of Natural History Vienna], 1995.

Guzzini, Stefano. "Structural Power: The Limits of Neorealist Power Analysis." *International Organization* 47, no. 3 (1993): 443–78.

Ha, Mai-Lan. "The Role of Regional Institutions in Sustainable Development: A Review of the Mekong River Commission's First 15 Years." *Consilience: The Journal of Sustainable Development* 5, no. 1 (2011): 125–40.

Haas, Ernst B. "Why Collaborate? Issue-Linkage and International Regimes." *World Politics* 32, no. 3 (1980): 357–405.

Hadfield, Peter. "Burmese villagers exiled from ancestral home as fate of dam remains unclear." *The Guardian*, 04.03.2014. Available online at www.theguardian.com/environment/2014/mar/04/burma-village-myitsone-dam-project-china (accessed 30.03.2014).

Haefner, Andrea. "Non-Traditional Security – The Case of Water Security in the Mekong Subregion." *Pacific Geographies* 40, July/August (2013a): 17–22.

—. "Regional Environmental Security: Cooperation and Challenges in the Mekong Subregion." *Global Change, Peace & Security* 25, no. 1 (2013b): 27–42.

—. "Cooperation and Conflict Along Transnational Rivers: The Role of Regional Institutions." In *Presentation at the Oceanic Conference for International Studies*. 1–17. Melbourne, July, 2014.

Haftendorn, Helga. "Water and International Conflict." *Third World Quarterly* 21, no. 1 (2000): 51–68.

Hamilton, Stephen K. "Potential Effects of a Major Navigation Project (Paraguay–Parana Hidrovia) on Inundation in the Pantanal Floodplains." *Regulated Rivers: Research & Management* 15, no. 4 (1999): 289–99.

Hammer, Jesse H., and Aaron T. Wolf. "Patterns in International Water Resource Treaties: The Transboundary Freshwater Dispute Database." *Colorado Journal of International Environmental Law and Policy* 9 (1998): 157–77.

Hanratty, Dannin M., and Sandra W. Meditz. *Paraguay: A Case Study – Itaipu, Yacreta, and Corpus*. Washington, DC: Washington Library of the Congress, 1988.

Harper, Krista. "'Wild Capitalism' and 'Ecocolonialism': A Tale of Two Rivers." *American Anthropologist* 107, no. 2 (2005): 221–33.

Haskel, Guillermo. "Paraguay Calls on Neighbours to Revamp Hidrovia Fleet and Ports." *Buenos Aires Herald*, 12.05.2014. Available online at www.buenosairesherald.com/article/159221/paraguay-calls-on-neighbours-to-revamp-hidrov%C3%ADa-fleet-and-ports (accessed 10.12.2014).

Hensengerth, Oliver. "Vietnam's Security Objectives in Mekong Basin Governance." *Journal of Vietnamese Studies* 3, no. 2 (2008): 101–27.

—. "Mekong Basin Cooperation – Current Development and Institutional Arrangements." In *Regionalism in China-Vietnam Relations: Institution-Building in the Greater Mekong Subregion*, edited by Oliver Hensengerth, 75–97. Florence, KY: Routledge, 2009a.

—. "Money and Security: China's Strategic Interests in the Mekong River Basin." *Chatman House*, no. 2 (2009b): 1–11.

—. "Transboundary River Cooperation and the Regional Public Good: The Case of the Mekong River." *Contemporary Southeast Asia* 31, no. 2 (2009c): 326–49.

Higham, Charles. *The Civilization of Angkor*. London: Weidenfeld and Nicholson, 2001.

Hillman, Robert. *Australians at War on the World Stage*. Melbourne: Echidna Books, 2006.

Hirsch, Philip. "Coming Full Circle: New Trends, Same Old Plans." *Watershed* 12, no. 3 (2008): 38–43.

—, Kurt Morck Jensen, with Ben Boer, Naomi Carrard, Stephen FitzGerald, and Rosemary Lyster. "National Interests and Transboundary Water Governance in the Mekong." Sydney: University of Sydney, 2006.

Hobbes, Thomas. *Leviathan*. Harmondsworth: Penguin Classics, 1986.

Hochstetler, Kathryn. "After the Boomerang: Environmental Movements and Politics in the La Plata River Basin." *Global Environmental Politics* 2, no. 4 (2002): 35–57.

Homer-Dixon, Thomas F. "Environmental Scarcities and Violent Conflict: Evidence from Cases." *International Security* 19, no. 1 (1994): 5–40.

Hook, Jacob, Susan Novak, and Robyn Johnston. "Social Atlas of the Lower Mekong Basin." 1–154. Phnom Penh: MRC, 2003.

Howley, Jessica. "The Gabcikovo–Nagymaros Case: The Influence of the International Court of Justice on the Law of Sustainable Development." *Queensland Law Student Review* 2, no. 1 (2009): 1–19.

Hsing, Loh Su. "China Holds the Key to Asia's 'Blue Gold'." *Jakarta Globe*, 15.12.2011.

Huang, Yan. "Uruguay." World Small Hydropower Development Report 2013, edited by Heng Liu and Diego Masera. United Nations Industrial Development Organization; International Center on Small Hydro Power, 2013.

Hudson, Chris. "The Role of International Environmental Law in the Protection of the Danube River Basin: The Baia Mare Cyanide Spill." *Colorado Journal of International Environmental Law and Policy* 12, no. 1 (2001): 16, 367–93.

Hungarian Ministry of Foreign Affairs. "Information on the Accidental Pollution and Related Mitigation Measures of the 'Red Mud Spill at Ajka'." Budapest: Hungarian Ministry of Foreign Affairs, 2010.

Hurrell, Andrew. "Security in Latin America." *International Affairs* 74, no. 3 (1998): 529–46.

Huxley, Tim. "Southeast Asia in 2007: Domestic Concerns, Delicate Bilateral Relations, and Patchy Regionalism." *Southeast Asian Affairs* (2008): 3–28.

Ikenberry, John, and Michael Mastanduno. "International Relations Theory and the Search for Regional Stability." In *International Relations Theory and the Asia–Pacific*, edited by John Ikenberry and Michael Mastanduno, 1–22. New York: Columbia University Press, 2003.

International Association of Danube Research (IAD). "International Association for Danube Research – Home." Available online at www.iad.gs/index.php?item=home (accessed 22.06.2012).

International Center for Environmental Management. "Environmental Assessment of Hydropower on the Mekong Mainstream – Summary of the Final Report." 1–23. Hanoi: International Center for Environmental Management, 2010.

International Commission for the Protection of the Danube River (ICPDR). "Convention on Cooperation for the Protection and Sustainable Use of the Danube River (Danube River Protection Convention)." Sofia, Bulgaria: ICPDR, 1994.

—. "Regional Inventory of Potential Accidental Risk Spots in the Tisa Catchment Area of Romania, Hungary, Ukraine and Slovakia." 1–25. Vienna: ICPDR in cooperation with Zinke Environment Consulting for Central and Eastern Europe, 2000.

—. "The Bystroe Canal: Dredging up Many Questions." Vienna: ICPDR, 2004a.

—. "Re: Letter Expressing Concerns About Ukraine Violation of the Convention on Cooperation for the Protection and Sustainable Use of the Danube River." Vienna: ICPDR, 2004b.

—. "Danube Facts and Figures – Austria." 1–6. Vienna: ICPDR, 2006a.

—. "Danube Facts and Figures – Hungary." 1–5. Vienna: ICPDR, 2006b.

—. "Danube Facts and Figures – Romania." 1–7. Vienna: ICPDR, 2006c.

—. "Danube Facts and Figures – Serbia." 1–6. Vienna: ICPDR, 2006d.

—. "Danube Facts and Figures – Bosnia and Herzegovina." 1–4. Vienna: ICPDR, 2007a.

—. "Danube Facts and Figures – Germany." 1–6. Vienna: ICPDR, 2007b.

—. "Danube Facts and Figures – Republic of Moldova." 1–3. Vienna: ICPDR, 2007c.

—. "Danube Facts and Figures – The Czech Republic." 1–5. Vienna: ICPDR, 2007d.

—. "Danube Facts and Figures – The Slovak Republic." 1–7. Vienna: ICPDR, 2007e.

—. "Danube Facts and Figures – Ukraine." 1–4. Vienna: ICPDR, 2007f.

—. "The Danube River Basin – Facts and Figures." 1–26. Vienna: ICPDR, 2009.

—. "Danube Facts and Figures." 1–13. Vienna: ICPDR, 2010a.

—. "Danube Facts and Figures – Croatia." 1–5. Vienna: ICPDR, 2010b.

—. "Danube Facts and Figures – Montenegro." 1–6. Vienna: ICPDR, 2010c.

—. "ICPDR – About Us." Available online at www.icpdr.org/icpdr-pages/about_us.htm (accessed 04.06.2012).

—. "ICPDR – Contracting Parties." Available online at www.icpdr.org/icpdr-pages/contracting_parties.htm (accessed 11.01 2012).
—. "ICPDR – Countries of the Danube." Available online at www.icpdr.org/icpdr-pages/countries.htm (accessed 11.01.2012).
—. "Permanent Secretariat of the ICPDR." Available online at www.icpdr.org/icpdr-pages/secretariat.htm (accessed 04.06.2012).
—. "Dams & Structures." Available online at www.icpdr.org/main/issues/dams-structures (accessed 13.11.2013).
—. "Organigram of the International Commission for the Protection of the Danube River (ICPDR)." Vienna: ICPDR, 2014.
International Court of Justice (ICJ). "Case Concerning Pulp Mills on the River Uruguay." 1–191. The Hague: ICJ, 2010a.
—. "Press Release: Pulp Mills on the River Uruguay (Argentina V. Uruguay)." 1–7. The Hague: ICJ, 2010b.
International Rivers Network. "Planned Dams in Vietnam." International Rivers Network, 2001.
—. "Navigation Project Threatens Livelihoods, Ecosystem." Berkeley, CA: International Rivers Network, 2002.
—. "The Xayaburi Dam – a Looming Threat to Mekong River." Berkeley, CA: International Rivers Network, 2011.
—. "Lower Sesan 2 Dam." Available online at www.internationalrivers.org/campaigns/lower-sesan-2-dam (accessed 23.05.2016).
—. "Poeyry's Role in the Xayabouri Dam Controversy." Bangkok: International Rivers Network, 2013a.
—. "The Mekong Dams Dispute: Four Trends to Watch." Vienna: International Rivers Network, 2013b.
—. "Paraguay–Parana Hidrovia." Available online at www.internationalrivers.org/campaigns/paraguay-paran%C3%A1-hidrovia (accessed 10.08.2014).
International Union for Conservation of Nature (IUCN). "Upper Mekong Development." Available online at http://archive-org.com/org/w/waterandnature.org/2013-11-20_3196779_28/Huong_Vietnam_E_flows_pilot_WANI/ (accessed 29.08.2013).
Jaehyon, Lee. "Prospects for Korean–Southeast Asian Relations." *Southeast Asia's Security and Political Outlook* (2010): 17–19.
Japan Environmental Council. "The Mekong Region: Incorporating the Views of Regional Civil Society." In *The State of the Environment in Asia*, edited by Japan Environmental Council, 131–50. Tokyo: Springer, 2005.
Jiang, Yong. "China's Water Scarcity." *Journal of Environmental Management* 90 (2009): 3185–96.
Junk, Wolfgang. "Current State of Knowledge Regarding South America Wetlands and Their Future under Global Climate Change." *Aquatic Science* 75, no. 1 (2013): 113–31.
Kempkey, Natalie, Margaret Pinard, Victor Pochat, and Ariel Dinar. "Negotiations over Water and Other Natural Resources in the La Plata River Basin: A Model for Other Transboundary Basins?." *International Negotiation* 14, no. 2 (2009): 253–79.
Keohane, Robert. *After Hegemony*. Princeton, NJ: Princeton University Press, 1984.
—, and Joseph Nye. *Power and Interdependence: World Politics in Transition*. Boston, MA: Little Brown, 1989.
—. "Interdependence in World Politics." In *The Theoretical Evolution of International Political Economy: A Reader*, edited by George T. Crane and Abla Amawi, 3–37. New York: Oxford University Press, 1997.

Klom, Andy. "Mercosur and Brazil: A European Perspective." *International Affairs* 79, no. 2 (2003): 351–68.
Koh, Kheng L., and Nicholas A. Robinson. "Regional Environmental Governance: Examining the Association of Southeast Asian Nations (ASEAN) Model." In *Global Environmental Governance*, edited by Daniel C. Esty and Maria H. Ivanova, 101–20. New Haven, CT: Yale School of Forestry and Environmental Studies, 2002.
Kongrut, Anchalee. "China Should Set Good Example on the Mekong River." *Economist Observer*, 25.01.2013. Available online at www.internationalrivers.org/resources/china-should-set-good-example-on-the-mekong-river-7806 (accessed 30.03.2013).
Koutsoukis, A J. *Australia and World Affairs 1850s–1980s*. Melbourne: Longman Cheshire, 1987.
Koyano, Mari. "The Significance of the Convention on Environmental Impact Assessment in a Transboundary Context (Espoo Convention) in International Environmental Law: Examining the Implications of the Danube Delta Case." *Impact Assessment and Project Appraisal* 26, no. 4 (2008): 299–314.
—. "Effective Implementation of International Environmental Agreements: Learning Lessons from the Danube Delta." In *Public Interest Rules of International Law*, edited by Teruo Komori and Karel Wellens, 9, 259–91. Farnham: Ashgate Publishing Group, 2009.
Kramer, John M. "EU Enlargement and the Environment: Six Challenges." *Environmental Politics* 13, no. 1 (2004): 290–311.
Krasner, Stephen D. *Structural Conflict: The Third World against Global Liberalism*. Berkeley, CA: University of California Press, 1985.
Krasno, Jean. "Brazil, Argentina Make It Official." *Bulletin of the Atomic Scientists* 48, no. 3 (1992): 10.
Krepper, C. M., and V. Venturini. "Assessing Interannual Water Balance of La Plata River Basin." *Atmosfera* 22, no. 4 (2009): 387–98.
Krongkaew, Medhi. "The Development of the Greater Mekong Subregion (GMS): Real Promise or False Hope?." *Journal of Asian Economics* 15, no. 5 (2004): 977–98.
Kuenzer, Claudia, Ian C. Campbell, Marthe Roch, Patrick Leinenkugel, Vo Quoc Tuan, and Stefan Dech. "Understanding the Impact of Hydropower Developments in the Context of Upstream-Downstream Relations in the Mekong River Basin." *Sustainable Science* 8, no. 4 (2012): 565–84.
Kuik, Cheng-Chwee. "Multilateralism in China's ASEAN Policy: Its Evolution, Characteristics, and Aspiration." *Contemporary Southeast Asia* 27, no. 1 (2005): 102–22.
Laboranti, Claudio. "Pilcomayo River Basin Institutional Structure." *International Journal of Water Resources Development* 27, no. 3 (2011): 539–54.
Laird, Sam. "Mercosur: Objectives and Achievements." Geneva: World Trade Organization (Trade Policy Review Division), 1997.
Lazarus, Kate, Pierre Dubeau, Channa Bambaradeniya, Richard Friend, and Latsamay Sylavong. *An Uncertain Future: Biodiversity and Livelihoods Along the Mekong River in Northern Lao PDR*. 1–49. Bangkok and Gland: The World Conservation Union (IUCN), 2006.
Le Breton, R. J. G. "Sustainable Land Use in the Atlantic Rainforest of Brazil." Iracambi, Brazil: Iracambi Atlantica Rainforest Research and Conservation Center, 1998.
Leech, Beth L. "Asking Questions: Techniques for Semistructured Interviews." *Political Science and Politics* 35, no. 4 (2002): 665–68.
Lerner, Michael A. "Dangerous Waters: Violations of International Law and Hydropower Development Along the Sesan River." 1–41. Phnom Penh: Oxfam America, 2003.

Lever, Lian. "Romania, Danube States, Agree to Keep River Open for Shipping." *Romania–Insider*, 07.06.2012. Available online at www.romania-insider.com/romania-danube-states-agree-to-keep-river-open-for-shipping/58525/ (accessed 12.6.2012).

Levy, Marc A., Peter M. Haas, and Robert O. Keohane. "Institutions for the Earth: Promoting International Environmental Protection." *Environment* 34, no. 4 (1992): 12–36.

Lima Barros Dolabella, Ana Lucia. "The Brazilian Pantanal: An Overview." In *The Pantanal of Brazil, Bolivia and Paraguay*, edited by Frederick A. Swarts, Chapter 3. Gouldsboro, PA: Hudson MacArthur Publishers, 2000.

Lowi, Miriam R. *Water and Power: The Politics of a Scarce Resource in the Jordan River Basin*. Cambridge: Cambridge University Press, 1993.

—. "Rivers of Conflict, Rivers of Peace." *Journal of International Affairs* 49, no. 1 (1995): 123–44.

—. "Water and Conflict in the Middle East and South Asia: Are Environmental Issues and Security Issues Linked?." *Journal of Environment and Development* 8, no. 4 (1999): 376–96.

Lukes, Steven. *Power: A Radical View*. Hong Kong: MacMillan Press, 1974.

McCaffrey, Stephen. "The Danube River Basin." In *The Multi-Governance of Water*, edited by Matthias Finger, Ludivine Tamiotti and Jeremy Allouche, 79–102. New York: State University of New York Press, 2006.

McCormick, John. *Understanding the European Union – A Concise Introduction*. 3rd ed. Hampshire: Palgrave Macmillan, 2005.

McDougall, Derek. "International Relations." In *Understanding Contemporary Asia Pacific*, edited by Katherine P. Kaup, 151–200. Boulder, CO: Lynne Rienner, 2007.

Machain, Andrea. "Why Brazil Gave Way on Itaipu Dam." *BBC Online*, 26.07.2009. Available online at http://news.bbc.co.uk/2/hi/8169139.stm (accessed 23.6.2014).

Makim, Abigail. "Resources for Security and Stability? The Politics of Regional Cooperation on the Mekong, 1957–2001." *Journal of Environment and Development* 11, no. 1 (2002): 5–52.

Martinez, Maria Eugenia, Martina Bonadeo, and Matias Barros. "Hidrovia Parana–Paraguay; Mercosur Y El Comercio [the Paraguay–Parana Waterway; Mercosur and Trading]." 1–10. Available online at www.slideshare.net/Jenny1994/hidrovia-paranparaguay-mara-eugenia-martnez-martina-bonadeo-y-matas-barrios (accessed 10.08.2014).

Marvel Shipping. "Deep Water Fairway (DWF) Danube – Black Sea, So Called 'Bystroe Canal'." Available online at www.gsh.delta-pilot.ua/index.php?mode=danube_presentation&lng=en (accessed 06 01.2014).

Mastny, Lisa. "Messing with the Mekong." 21–8. *World Watch*, November/December, 2003.

Mearsheimer, John J. "The False Promise of International Institutions." *International Security* 19, no. 3 (1995): 5–49.

—. *The Tragedy of Great Power Politics*. New York: W. W. Norton & Company, 2001.

Mecham, Michael. "Mercosur: A Failing Development Project?." *International Affairs* 79, no. 2 (2003): 369–87.

Mehtonen, Katri, Marko Keskinen, and Olli Varis. "The Mekong: IWRM and Institutions." In *Management of Transboundary Rivers and Lakes*, edited by Olli Varis, Asti K. Biswas and Cecilia Tortajada, 207–26. Berlin: Springer, 2008.

Mekong River Commission (MRC). "Agreement on the Cooperation for the Sustainable Development of the Mekong River Basin." Chiang Rai, Thailand: MRC, 1995.

—. "Annual Report 2001 – Mekong River Commission." 1–24. Vientiane: MRC, 2002.
—. "Mekong River Commission – Annual Report 2004." Vientiane: MRC, 2005.
—. "Mekong Prime Ministers Agree to Prioritise Climate Change as Summit Ends." Available online at www.mrcmekong.org/news-and-events/news/mekong-prime-ministers-agree-to-prioritise-climate-change-as-summit-ends/ (accessed 23.05.2016).
—. "Mekong River Commission – Audited Financial Statements." 1–14. Hanoi: MRC, 2012.
—. "MRC Appoints New Chief Executive Officer." Available online at www.mrcmekong.org/news-and-events/news/mrc-appoints-new-chief-executive-officer/ (accessed 05.07.2012).
—. "MRC Council concludes a future direction for the organisation", 16.01.2016, Phnom Penh, Cambodia. Available online at www.mrcmekong.org/news-and-events/news/mrc-council-concludes-a-future-direction-for-the-organisation/ (accessed 25.01.2016).
—. "Sweden Provides US $8.5 Million to Strengthen Sustainable Development and Management of Mekong Resources." Available online at www.mrcmekong.org/news-and-events/news/sweden-provides-us-8-5-million-to-strengthen-sustainable-development-and-management-of-mekong-resources/ (accessed 25.07.2012).
—. "About the MRC." Available online at www.mrcmekong.org/about-mrc/ (accessed 23.05.2016).
—. "MRC Financial Statements and Independent Auditor's Report Year Ended 31 December 2014." 1–18. Vientiane: MRC, 2015.
*Mekong Watch*. "Lower Sesan 2 Hydropower Project, Northeastern Cambodia." Available online at www.mekongwatch.org/english/country/cambodia/LS2/index.html (accessed 01.07.2015).
Mercosur. "Protocolo De Olivos [Olivos Protocol]." 1–22. Olivos, Argentina: Mercosur, 2002.
—. "Protocolo De Ouro Preto [Ouro Preto Protocol]." Available online at www.mercosur.int/innovaportal/file/721/1/cmc_1994_protocolo_ouro_preto_es.pdf (accessed 09.10.2012).
—. "Quienes Somos [Who We Are]." Available online at www.mercosur.int/innovaportal/v/6283/2/innova.front/quienes-somos (accessed 09.10.2012).
Mihic, Svetlana, Mirjana Golusin, and Milan Mihajlovic. "Policy and Promotion of Sustainable Inland Waterway Transport in Europe – Danube River." *Renewable and Sustainable Energy Reviews* 15 (2011): 1801–9.
Miklos, Domokos. "History and Results of the Hydrological Co-Operation of the Countries Sharing the Danube Catchment (1971–2008)." In *Hydrological Processes of the Danube River Basin*, edited by Mitja Brilly, 1–23. Dordrecht: Springer Science and Business Media, 2010.
—, and Jeno Sass. "The Danube River and its Basin Physical Characteristics, Water Regime and Water Balance." In *Hydrological Processes of the Danube River Basin*, edited by Mitja Brilly, 25–77. Dordrecht: Springer Science and Business Media, 2010.
Milich, Lenard, and Robert G. Varady. "Openness, Sustainability, and Public Participation: New Designs for Transboundary River Basin Institutions." *The Journal of Environmental Development* 8, no. 3 (1999): 258–306.
Mills, Don. "Romania Sues Australian Miner after Cyanide Spill." *National Post*, 12.02.2000. Available online at www.abc.net.au/science/articles/2000/07/11/150431.htm (accessed 12.09.2013).

Mirumachi, Naho. "Transboundary Water Security: Reviewing the Importance of National Regulatory and Accountability Capacities in International Transboundary River Basins." 1–24. Department of Geography, King's College London Working Paper No. 64, 2013.

—. *Transboundary Water Politics in the Developing World*. New York: Routledge, 2015.

Mitchell, David. "Proposed Danube Dam at Hainburg: WWF Protest and Upheld Injunction." *Environmental Conservation* 12, no. 1 (1985): 81.

Moellenkamp, Sabine. "The 'WFD–Effect' on Upstream-Downstream Relations in International River Basins – Insights from the Rhine and the Elbe Basins." *Hydrology and Earth System Sciences Discussions* 4 (2007): 1407–28.

Monahan, Jane. "Paraguay Looks to Stem Flow of Cheap Power from Itaipu Dam to Brazil." *Energy Economist*, no. 286 (2005): 14–16.

Morgenthau, Hans. *Politics Among Nations: The Struggle for Power and Peace*. 4th ed. New York: Knopf, 1948.

Motta Veiga, Pedro. "Brazil in Mercosur: Reciprocal Influence." In *Mercosur: Regional Integration, World Markets*, edited by Riordan Roett, 25–33. Boulder, CO: Lynne Rienner, 1999.

Myint, Tun. "Adaptability of International River Basin Regimes: Linkage Problems in the Rhine." In *Adaptive and Integrated Water Management*, edited by Claudia Pahl-Wostl, Pavel Kabat and Joern Moeltgen, 125–45. Berlin: Springer-Verlag, 2008.

—. *Governing International Rivers – Polycentric Politics in the Mekong and the Rhine*. Cheltenham: Edward Elgar, 2012.

Nam, So. "Fisheries Resources in Cambodia – Current Status, Key Issues and Directions." Available online at www.icem.com.au/documents/envassessment/mrc_sea_hp/1.%20inception/presentations/cambodia%20scoping%20workshop/pdf/Fisheries%20Resources%20in%20Cambodia%20-%20Current%20Status,%20Key%20Issues,%20Directions.pdf (accessed 03.11.2014).

National Audit Office Bulgaria. "Report on the Regularity and Performance Audit of the Application of the Provisions of the Danube River Protection Convention." 1–53. Sofia, Bulgaria: National Audit Office Bulgaria, 2001.

Nejamkis, Guido, and Ana Flor. "Mercosur Welcomes Venezuela, Suspends Paraguay." *Reuters.com*, 29.07.2012. Available online at www.reuters.com/article/2012/06/30/us-mercosur-idUSBRE85S1JT20120630 (accessed 04.10.2012).

Norman, Emma S. *Governing Transboundary Waters – Canada, the United States and Indigenous communities*. New York: Routledge, 2015.

Nye, Joseph S. *Soft Power*. New York: Public Affairs, 2004.

Organization of American States (OAS). "La Plata River Basin." *Water Projects Series*, no. 6, 2005a.

—. "La Plata River Basin – A Framework for the Sustainable Management of its Water Resources with Respect to the Hydrological Effects of Climate Variability and Change." *Water Projects Series*, 2005b.

—. "OAS – Who We Are." Available online at www.oas.org/en/about/who_we_are.asp (accessed 20.05.2014).

Osborne, Milton. *River Road to China. The Search for the Source of the Mekong*. Singapore: Archipelago, 1996.

—. "The Strategic Significance of the Mekong." *Contemporary Southeast Asia* 22, no. 3 (2000): 429–44.

—. *The Mekong: Turbulent Past, Uncertain Future*. 2nd ed. Crows Nest, Australia: Allen and Unwin, 2006.

Oxfam Australia. "Lower Sesan 2 Dam." Available online at www.oxfam.org.au/explore/infrastructure-people-and-environment/save-the-mekong/3s-critical-environmental-zone/lower-sesan-2-dam/ (accessed 10.08.2013).

—. "Xayaburi Dam." Available online at www.oxfam.org.au/what-we-do/infrastructure-people-and-environment/save-the-mekong/damming-the-mekong/xayaburi-dam/ (accessed 10.09.2014).

Payne, Cymie R. "Environmental Impact Assessment as Duty under International Law: The International Court of Justice Judgement on Pulp Mills on the River Uruguay." *The European Journal of Risk Regulation* 3 (2010): 317–24.

People's Republic of China. "White Paper on China's National Defense in 2002. Available online at www.fas.org/nuke/guide/china/doctrine/natdef2002.html (accessed 21.06.2011).

Petherick, Anna "Science Panel Gives Hope in River-Pollution Dispute." *Nature* 466, no. 911 (2010).

Petrovic, Pavel, Katarina Mravcova, Ladislav Holko, Zdenek Kostka, and Pavol Miklanek. "Basin-Wide Water Balance in the Danube River Basin." In *Hydrological Processes of the Danube River Basin*, edited by Mitja Brilly, 227–58, 683. Dordrecht: Springer Science and Business Media, 2010.

Phillips, David, Marwa Daoundy, Stephen McCaffrey, Joakim Ojendal, and Anthony Turton. "Trans-Boundary Water Co-Operation as a Tool for Conflict Prevention and Broader Benefit Sharing." Global Development Studies No. 4. Stockholm: Ministry of Foreign Affairs, 2006.

*Phys.org*. "Sunken Barge Threatens Oil Spill in Danube, WWF Says." Available online at http://phys.org/news/2012-11-sunken-barge-threatens-oil-danube.html (accessed 20.11.2013).

Pinder, John, and Simon Usherwood. *The European Union – A Very Short Introduction*. 2nd ed. Oxford: Oxford University Press, 2007.

Pochat, Victor. "Dams and Development Project: Identification, Collection of Information and Compilation of Relevant Practices Concerning the Integration into Policy/Normative Frameworks and Implementation of Key Priority Issue: International Policy in Shared River Basins." 96–103. Available online at www.unep.org/dams/files/Compendium/Report_InterRivers.pdf (accessed 10.10.2014).

—. "The La Plata River Basin." In *7th Biennial Rosenberg International Forum on Water Policy – Water for the Americas: Challenges and Opportunities*. 1–30. Buenos Aires, November, 2010.

—. "International Agreements, Institutions and Projects in La Plata River Basin." *International Journal of Water Resources Development* 27, no. 3 (2011): 497–510.

Ponce, Victor M. "Geographical Background, Pantanal and Hidrovia." In *Hidrovia Report – Wetland Hydrology*. San Diego, CA: San Diego State University, n.d.

Powell, Robert. "Bargaining Theory and International Conflict." *Annual Review of Political Science* 5 (2002): 1–30.

Priscoli, Jerome Delli, and Aaron T. Wolf. *Managing and Transforming Water Conflicts*. Cambridge: Cambridge University Press, 2009.

Pun, Santa B. "Paraguay, Bhutan and Nepal: Landlocked but Hydropower Rich." *Hydro Nepal* 3, no. 3 (2008): 1–5.

Pye, Daniel, and Khouth Sophak Chakrya. "Sesan dam gets new design." *The Phnom Penh Post*, 30.03.2015. Available online at www.phnompenhpost.com/national/sesan-dam-gets-new-design (accessed 05.04.2015).

Qin, Jize. "Summit to Enhance Co-Ops Along Mekong." *China Daily*, 07.05.2005, Available online at www.chinadaily.com.cn/english/doc/2005-07/05/content_457155.htm (accessed 13.09.2011).

Quiros, Rolando. "The Plata River Basin: International Basin Development and Riverine Fisheries." In *Proceedings of the Second International Symposium on the Management of Large Rivers for Fisheries Volume I*, edited by Ian G. Cowx, O. Almeida, C. Bene, R. Brummett, S. Bush, W. Darwall, J. Pittock and M. van Brakel, 253–71. Bangkok: FAO Regional Office for Asia and the Pacific, 2004.

Radio Free Asia. "Laos Confirms 'Preparatory' Work on Controversial Dam Project." Available online at www.rfa.org/english/news/laos/confirms-07312013151754.html (accessed 10.08.2013).

Ramsar Convention on Wetlands. "Bystroe Canal Project under International Scrutiny." Available online at www.ramsar.org/news/bystroe-canal-project-under-international-crutiny (accessed 13.04.2014).

Ravenhill, John. "Fighting Irrelevance: An Economic Community 'with ASEAN Characteristics'." *The Pacific Review* 21, no. 4 (2008): 469–88.

Rensmann, Thilo. "Recent Development in the Jurisprudence of the International Court of Justice." 1–6. Dresden: TU Dresden, 2010.

Resende-Santos, Joao. "The Origins of Security Cooperation in the Southern Cone." *Latin American Politics and Society* 44, no. 4 (2002): 89–126.

Roberts, Tyson R. "Downstream Ecological Implications of China's Lancang Hydropower and Mekong Navigation Project." In International Symposium on Biodiversity Management and Sustainable Development in the Lancang-Mekong River Basin. Xishuangbanna, China, December, 2001.

Rubinstein, Ariel. "Perfect Equilibrium in a Bargaining Model." *Econometrica* 50, no. 1 (1982): 97–109.

Rucks, Jorge. "Experience and Role of the Organisation of American States in Transboundary River Basin Management in Latin America." In *Climate and Water – Transboundary Challenges in the Americas*, edited by Henry F. Diaz and Barbara J. Morehouse, 41–58. Dordrecht: Kluwer Academic Publishers, 2003.

Sadoff, Claudia W., and David Grey. "Beyond the River: The Benefits of Cooperation on International Rivers." *Water Policy* 4, no. 5 (2002): 389–403.

Schelling, Thomas C. *The Strategy of Conflict*. Harvard, MA: Harvard University Press, 1980.

Schmeier, Susanne. *Governing International Watercourses – River Basin Organizations and the Sustainable Governance of Internationally Shared Rivers and Lakes*. Abingdon: Routledge, 2013a.

—. "Hydropower Development in the LMB – Updated June 2013." 1–6. Vientiane: MRC–GIZ Cooperation Program, 2013b.

—. "The Danube River Basin and the ICPDR: Strong Achievements by a Narrow Institution." In *Governing International Watercourses – River Basin Organizations and the Sustainable Governance of Internationally Shared Rivers and Lakes*, 171–216. Abingdon: Routledge, 2013c.

—, Andrea K. Gerlak, and Sabine Schulze. "Who Governs Internationally Shared Watercourses? Clearing the Muddy Waters of International River Basin Organizations." Earth System Governance Working Paper No. 8, 2013.

Schneider, Keith. "Fierce Competition over Water Threatens China's Economic Progress and Global Food, Energy Prices." *Yale Global*, 12.07.2011. Available online at http://yaleglobal.yale.edu/content/water-challenges-asia-powers-part-i (accessed 13.01.2014).

Schreurs, Miranda. "Environmental Protection in an Expanding European Community: Lessons from Past Accessions." *Environmental Politics* 13, no. 1 (2004): 25–51.

Schultz, Bart, and Stefan Uhlenbrook. "Water Security: What Does It Mean, What May It Imply?." Delft: UNESCO–IHE (Institute for Water Education), 2007.

Southeast Asia River Network. "Downstream Impacts of Hydropower and Development of an International River: A Case Study of Lancang–Mekong." 1–5. Chiang Mai: Southeast Asia River Network, 2004.

Seawright, Jason, and John Gerring. "Case Selection Techniques in Case Study Research." *Political Research Quarterly* 61, no. 2 (2008): 294–308.

Sen, David, and Vong Sokheng. "Questions Follow on Heels of Hydropower Dam Sesan 2 Vote." *The Phnom Penh Post*, 18.02.2013. Available online at www.phnompenhpost.com/national/questions-follow-heels-hydropower-dam-sesan-2-vote (retrieved 25.02.2013).

Severino, Rodolfo C. "ASEAN: New Charter, New Optimism." *Southeast Asia Security Outlook* 153, 2009/2010 (2010): 7–9.

Shepherd, Kirstie. "Work Resumes on Bystroe Canal." *Danube Watch* 3/2006.

—. "Finding a Common Ground for International Cooperation." *Danube Watch*, 2/2008.

Simon, Sheldon W. "ASEAN and the New Regional Multilateralism." In *International Relations of Asia*, edited by David Shambaugh and Michael Yahuda, 195–214. Lanham, MD: Rowan & Littlefield Publishers, 2008.

Smith, Scott E., Ferenc Szilagyi, and Lajos Horvath. "Environmental Impacts of the Gabcikovo Barrage System to the Szigetkoez Region." *Clean Technologies and Environmental Policy* 4, no. 1 (2002): 54–64.

Snyder, Jack. "One World, Rival Theories." *Foreign Policy Journal* November/December, no. 145 (2004): 52–63.

Söderbaum, Frederik. "Introduction: Theories of New Regionalism." In *Theories of New Regionalism*, edited by Frederik Söderbaum and Timothy M. Shaw, 1–21. Hampshire and New York: Palgrave Macmillan, 2003.

Soldan, Premysl, Michal Pavonic, Jan Boucek, and Jiri Kokes. "Baia Mare Accident – Brief Ecotoxiological Report of Czech Experts." *Ecotoxicology and Environmental Safety* 49, no. 3 (2001): 255–61.

Staras, Micrea. "Documentation on the Likely Significant Transboundary Impact of the Ukrainian Deep-Water Navigation Canal Danube Black Sea in the Context of Espoo Convention, 1991." 1–34. Tulcea, Romania: Ministry of Environment and Water Management, Danube Delta National Institute for Research and Development, 2005.

Stark, Jeffrey. "Environmental Security in Latin America and the Caribbean: Development, Conflict, and Democracy." In *New and Critical Security and Regionalism – Beyond the Nation State*, edited by James J. Hentz and Morten Boas, 55–74. Hampshire: Ashgate Publishing Limited, 2003.

Stec, Stephen, Alexios Antypas, Tonya Jansen, and Eszter Gulacsy. "Transboundary Environmental Governance and the Baia Mare Cyanide Spill." *Review of Central and East European Law* 27, no. 4 (2001): 639–91.

Stuart-Fox, Martin. "Southeast Asia and China: The Role of History and Culture in Shaping Future Relations." *Contemporary Southeast Asia* 26, no. 1 (2004): 116–38.

Stubbs, Richard. "Meeting the Challenges of Region-Building in ASEAN." In *Contemporary Southeast Asia*, edited by Mark Beeson, 234–52. Hampshire: Palgrave Macmillan, 2009.

Sunchindah, Apichai. "The ASEAN Approach to Regional Environmental Management." Jakarta, Indonesia: ASEAN Secretariat, 2002.

Tan, Kenneth P. "Crisis, Self-Reflection, and Rebirth in Singapore's National Life Cycle." *Southeast Asian Affairs* (2003): 241–58.

Targeted News Services. "13th ASEAN–China Summit." *Targeted News Services*, 30.10.2010. Available online at http://targetednews.com/ (accessed 30.05.2013).

Task Force on Transboundary Waters. "Transboundary Waters: Sharing Benefits, Sharing Responsibilities." Zaragoza, Spain: UNESCO, 2008.

Than, Mya, and George Abonyi. "The Greater Mekong Subregion: Co-Operation in Infrastructure and Finance." In *ASEAN Enlargement: Impacts and Implications*, edited by Mya Than and Carolyn Gates, 128–63. Singapore: Institute of Southeast Asian Studies, 2001.

The Economist Intelligence Unit. "Brazil/Paraguay Politics: A Dam Deal." *EIU Views Wire*, 27.07.2009.

Timerman, Hector. "Our Alliance with Brazil Is Indestructible." *Buenos Aires Herald*, 30.01.2011. Available online at www.buenosairesherald.com/article/57697/our-alliance-with-brazil-is-indestructible-timerman-says (accessed 09.10.2012).

Treaty of the Financial Fund for the Development of the La Plata Basin (FONPLATA). "Convenio Constitutivo Del Fondo Financiero Para El Desarrollo De La Cuenca Del Plata [Agreement of the Financial Fund for the Development of the La Plata Basin]." 1–13. Buenos Aires: FONPLATA, 1974.

—. "Memoria Institucional Fonplata 2010 [2010 Annual Report of the Fonplata Institution]." 1–94. Santa Cruz de la Sierra, Bolivia: FONPLATA, 2011.

—. "Fondo Financiero Para El Desarrollo De La Cuenca Del Plata [Financial Fund for the Development of the La Plata Basin]." Available online at www.fonplata.org/ (accessed 04.10.2012).

—. "Estructura del Capital [Structure of the Capital]." Available online at www.fonplata.org/index.php/fonplata/nuestra-historia (accessed 21.04.2015).

Tucci, Carlos E. M., and Robin T. Clarke. "Environmental Issues in the La Plata Basin." *International Journal of Water Resources* 14, no. 2 (1998): 157–73.

*Ultima Hora*. "Argentina Y Paraguay Acuerdan Iniciar Renegociacion Del Tratado De Yacyreta [Argentina and Paraguay Agree to Initiate Renegotiations of the Yacyreta Treaty]," 12. *Ultima Hora*, 22.04.2014.

United Nations (UN). "Convention on the Protection and Use of Transboundary Watercourses and International Lakes." Helsinki: UN, 1992.

—. "Convention on the Law of the Non-Navigational Uses of International Watercourses." Available online at https://treaties.un.org/Pages/ViewDetails.aspx?src=IND&mtdsg_no=XXVII-12&chapter=27&lang=en (accessed 13.10.2014).

United Nations Economic Commission for Europe (UNECE). "Bystroe Canal Project and Convention on Environmental Impact Assessment in a Transboundary Context (Espoo, 1991)." Available online at www.osce.org/eea/32889?download=true (accessed 01.11.2014).

United Nations Educational, Scientific and Cultural Organisation (UNESCO). "United Nations World Water Development Report (WWDR 3): Water in a Changing World." Paris and London: UNESCO, 2009.

—. "IHP Danube Cooperation." Available online at www.unesco.org/new/en/venice/natural-sciences/water/danube-cooperation/ (accessed 07.06.2012).

—. "Principles." Available online at www.unesco.org/new/fileadmin/MULTIMEDIA/FIELD/Venice/pdf/NS/Principles.pdf (accessed 07.06.2012).

United Nations Environmental Program (UNEP). "Negotiating and Implementing MEAS: A Manual for NGOs." 137–9. UNEP, 2007.

United Nations Environmental Program (UNEP)/GRID–Arendal. "Balkan Vital Graphics – Environment without Borders." Zemun, Serbia: UNEP/GRID–Arendal, 2007.

United Nations Environmental Program (UNEP)/UN Office for the Co-ordination of Humanitarian Affairs (OCHA). "UNEP/OCHA Assessment Mission – Cyanide Spill at Baia Mare Report." Edited by UNEP/OCHA, 1–57. UNEO/OCHA: Geneva, 2000.

United Nations Watercourses Convention. "Article 33." Available online at http://legal.un.org/ilc/texts/instruments/english/conventions/8_3_1997.pdf (accessed 01.03.2014).

Varchol, Douglas. *Movie: The Mekong*. CGIAR, IUCN's Mekong Water Dialogue and the Swedish International Development Agency, 2012.

*Wall Street Journal*. "Tension over Dam Project Shifts Myanmar's Politics." *The Wall Street Journal*, 13.08.2011. Available online at http://online.wsj.com/article/SB1000 1424053111904006104576504123439749638.html (accessed 04.10.2013).

Waltz, Kenneth N. *Theory of International Politics*. New York: Random House, 1979.

Weber, Max. *The Theory of Social and Economic Organization*. Chicago, IL: Free Press, 1947.

Wolf, Aaron T. "Conflict and Cooperation Along International Waterways." *Water Policy* 1, no. 2 (1998): 251–65.

—, and Joshua T. Newton. "Case Study of Transboundary Dispute Resolution: The Environmental Program for the Danube River." 1–11. Corvallis, OR: College of Science and Institute for Water and Watersheds, Oregon State University, n.d.-a.

—. "Case Study Transboundary Dispute Resolution: The La Plata Basin." Institute for Water and Watersheds, n.d.-b.

World Water Assessment Program. "The United Nations World Water Development Report 3 – Case Study Volume: Facing the Challenges." Available online at http://webworld.unesco.org/water/wwap/wwdr/wwdr3/case_studies/pdf/Case_Studies_LatinAmCaribbean.pdf#page=2 (accessed 05.01.2012).

World Wide Fund for Nature (WWF). "World's Top 10 Rivers at Risk." 1–53. Gland: WWF, 2007.

—. "Bystroe Canal/Danube–Black Sea Deep Navigation Channel." 1–3. Vienna: WWF, 2009a.

—. "Save the Danube as a Lifeline! – Steps Towards Sustainable Navigation. Common NGO Position on Navigation in the Danube Basin." 1–14. Vienna: WWF, 2009b.

Worrell, Shane. "Dam Firm Violated Law." *The Phnom Penh Post*, 05.06.2013. Available online at www.phnompenhpost.com/national/dam-firm-violated-law (accessed 25.06.2013).

Wunderlich, Jens-Uwe. "Regionalism and Integration Theory. The First Wave: Traditional Approaches." In *Regionalism, Globalisation and International Order*, edited by Jens-Uwe Wunderlich, 7–28. Hampshire and Burlington: Ashgate, 2007.

*Xinhua News Agency*. "Mekong Countries Agree on Commercial Navigation Pact." *Xinhua News Agency*, 18.04.2000. Available online at https://journal.probeinternational.org/2000/04/20/mekong-river-pact-be-signed-today/ (accessed 03.09.2013).

—. "Myanmar to Host Four-Country Commercial Navigation Meeting." *Xinhua News Agency*, 05.04.2007. Available online at http://en.people.cn/200704/05/eng20070405_363989.html (accessed 03.09.2013).

Yin, Robert K. *Case Study Research: Design and Methods*. Thousand Oaks, CA: SAGE, 1989.

Young, Oran R. "The Politics of International Regime Formation: Managing Natural Resources and the Environment." *International Organization* 43, no. 3 (1989): 349–75.

Zeitoun, Mark, and J. A. Allan. "Applying Hegemony and Power Theory to Transboundary Water Analysis." *Water Policy* 10, no. 2 (2008): 3–12, 41.

—, and Naho Mirumachi. "Transboundary Water Interaction I: Reconsidering Conflict and Cooperation." *International Environmental Agreements* 8, December (2008): 297–316.

Zibechi, Raul. "Is Brazil Creating its Own 'Backyard'?." Washington, DC: Center for International Policy, 2009.

Ziv, Guy, Eric Baranb, So Namc, Ignacio Rodríguez, and Simon A. Levina. "Trading-Off Fish Biodiversity, Food Security, and Hydropower in the Mekong River Basin." *Proceedings of the National Academy of Science of United States of America* 109, no. 15 (2012): 1–6.

# Index

For Product Safety Concerns and Information please contact our EU representative GPSR@taylorandfrancis.com
Taylor & Francis Verlag GmbH, Kaufingerstraße 24, 80331 München, Germany

www.ingramcontent.com/pod-product-compliance
Ingram Content Group UK Ltd.
Pitfield, Milton Keynes, MK11 3LW, UK
UKHW022119190926
475702UK00013B/1533
*9780367000257*